Eighteen Months on the Toilet and Other Feats, Facts and Astonishing Stats

The Best of Number Crunch Volume 1: Revised and Updated

John S. Croucher
Professor of Statistics Macquarie University
Cartoons by Roger Harvey

16

EasyRead Large

RHYW

Copyright Page from the Original Book

Woodslane Press Pty Ltd

7/5 Vuko Place, Warriewood, NSW 2102

Email: info@woodslane.com.au

Website: www.woodslane.com.au

Print edition first published in 2006 by Macmillan
Revised and updated edition published in 2010 by Woodslane Press
Ebook edition published by Woodslane Press in 2011

National Library of Australia Cataloguing-in-Publication entry

Author:	Croucher, John S.
Title:	Eighteen months on the toilet (electronic resource): and other feats, facts and astonishing stats: best of number crunch 1 / John S Croucher.
Edition:	2nd ed.
ISBN:	9781921683992 (ebook)
Subjects:	Statistics. Mathematical statistics. Newspapers—Sections, columns, etc.
Dewey Number:	519.5

Design and layout by Jenny Cowan Design

Set in 16 pt. Verdana

TABLE OF CONTENTS

Dedicated to my dear father

Sydney Ross Croucher
1920–2009

Preface

It is no secret that many people have a fascination with numbers – especially in this computer age. There is now the opportunity for recording and accessing more information than was ever dreamed possible by our ancestors. The seemingly endless stream of statistics quoted by commentators during sporting broadcasts is one example of our thirst for facts and figures; we digest them and draw our own conclusions on their meanings.

The genesis of this book was the response to my newspaper columns that appear around the world; the public's imagination has been captured by numbers both amazing and challenging. For over ten years the weekly 'Number Crunch' column in the *Sydney Morning Herald* and *The Age* in Melbourne have provided riveting data to Australia, while the weekly 'Statistically Speaking' column that appeared recently in the *Mail on Sunday* in the United Kingdom offered international readers the opportunity to be fascinated by statistics relevant to their own part of the world. These columns are read by millions of readers each week and provide testimony to just how much fun numbers can be.

For me the fascination with figures began very early. I have always been interested in numbers – for cricket, football and tennis especially but indeed for all sports. I pored over almanacs and sporting results. Numbers talked, sang, even shouted at me. I could see that there was so much to be achieved by analysing them. They told you how things were. They gave you a measure – a standard – against which other things, past, present and future, could be assessed.

This obsession has led to me being known as 'Professor FootyTAB' on television in Australia and 'Dr Number from Down Under' in my BBC Radio Quiz in Scotland, to my weekly columns in newspapers, and now to this book.

This edition of the best of *Number Crunch* includes over 300 updated entries, revisions and new items and also contains a collection of some of the more captivating statistics I have found. It is easy to dip into them, staring at any section. With over 2,500 facts and figures and around 150 matching humorous quotations, there is plenty of information to keep the reader absorbed and entertained. While there are specific categories, the 'Quirky' section in particular contains those that have established themselves as the most popular over the years.

The sources of these statistics are many and varied – ranging from encyclopaedias, books of facts, news reports, surveys and journal articles – all current at the time of writing. Those items that reveal the views and attributes of populations rely on survey data and are of course dependent on the particular respondents in the sample. But they are all from reliable samples of reasonable size and are often the figures that promote the most discussion among readers. A 'Further Reading' section is provided for those wishing to learn more.

This volume would not have been possible without the loving assistance of my wonderful wife Rosalind, without whose laughter, patience and guidance made it such an enjoyable project. She is still, and always will be, the light of my life.

JOHN S. CROUCHER
2010

QUIRKY

Well, if I called the wrong number, why did you answer the phone?

***James Thurber,* cartoon in the New Yorker**

Women's clothes: never wear anything that panics the cat.

P.J. O'Rourke,* Modern Manners, *1983

***

Average days for a plucked hair to reappear: **56**

Fastest time to calculate the square root of the six-digit number 732,915 without a calculator or computer (by M. Hari Prasad, India, in 1999): **63.8 seconds**
his answer:
856.1045496

Longest ear hair (Anthony Victor, India, 2007):
18.1 centimetres

World's heaviest woman ever (Rosalie Bradford, USA, 1987):
544 kilograms

Average number of sesame seeds reportedly on a Big Mac bun:
178

Proportion of British men who would prefer their football team win the premiership rather than having sex with the woman of their dreams: **40 per cent**

Number of times the word 'Bible' occurs in the works of Shakespeare: **0**

Number of witches in Shakespeare's Macbeth: **3**

Number of variations in the first four moves of a game of chess:
318,979,564,000

Length of the world's largest foot in 2009 (Sultan Kosen, Turkey):
36.5 centimetres
his hand length (wrist to fingertip):
27.5 centimetres
his height (also current world's tallest man): **246.5 centimetres**
his age: **26**

Annual cost worldwide to the aviation industry as a result of bird strikes: **$A3.5 billion**
number of lives lost since 1912: **190**

World's tallest living woman (Yao Defens, China):
236.2 centimetres

World's tallest living girl in 2010 (Malee Duangdee, Thailand):
208 centimetres
her age: **17 years**

Proportion of Americans who will not visit the dentist out of fear:
15 per cent

Truth is a rare and precious commodity. We must use it sparingly.

C.P. Scott,* Spectator, *1982

***

Youngest *Titanic* survivor (Millvina Dean, UK): **9 weeks**
number of her lifeboat: **13**

Number of years that Rip van Winkle slept: **20**

Longest time trapped in a lift (a 76-year-old woman in Cyprus, 1976): **6 days**

Price of a Model–T Ford when it was first produced in 1908:
$US850
number of colours it was available in between 1908 and 1926:
1 (black)

Usual number of beads in a Catholic Rosary:
between 38 and 169
an Anglican Rosary: **33**

Number of lines in the role of *Hamlet:* **1,422**
of words: **11,610**

Number of poker machines in Las Vegas:
about 150,000
Australia:
about 200,000
the world:
about 1 million

Proportion of Australians who say their sex life is monotonous:
6 per cent
Japanese:
13 per cent (highest)
Indians:
3 per cent (lowest)

Average amount spent each year on toys per child in the USA:
$A332
Africa: **$A2**
worldwide: **$A32**

Number of people in the world aged over 100 as at 1 November 2009:
450,000
percentage who are women:
85 per cent

Average number of nuts in a Snickers bar: **15**

Proportion of M&Ms that are brown: **30 per cent**

red: **20 per cent**
yellow: **20 per cent**

Largest recorded distance travelled by a single automobile
(1966 Volvo P1800 sports car):
3.4 million kilometres
age at which its engine was rebuilt:
1.1 million kilometres
its transmission replaced:
3.2 million kilometres

When I go to the beauty parlour I always use the emergency entrance.

Sometimes I just go for an estimate.

Phyllis Diller

***

Proportion of middle-aged Australian women who have unwanted facial hair:
25 per cent

Number of inanimate symbols in the zodiac:
1 (Libra – the scales)

Proportion of the world's countries that drive on the left:
30 per cent

Average life expectancy of a cat:
15 years
proportion of time it spends sleeping:
60 per cent

Number of cigarettes smoked annually worldwide:
about 5.3 trillion

Number of people in prison worldwide in 2010:
about 9 million

Number of horses involved in World War I:
about 1.5 million
proportion that were killed:
1 in 3

Estimated number of hits on the Google search engine each day:
about 2 billion

Number of domestic items mailed annually worldwide in 2006:
431 billion
international items:
5.6 billion

Proportion of occasions that people going on vacation forget to pack a toothbrush:
42 per cent
underwear:
22 per cent

Longest distance run on a treadmill in 24 hours (by Arulananthan Suresh Joachim, 2004):
257.88 kilometres

Number of pencils manufactured worldwide each year:
14 billion
average length they can write:
55 kilometres
number of words:

45,000

Average time a woman spends in a store if accompanied by another woman:
8 minutes 15 seconds
her children:
7 minutes 19 seconds
a man:
4 minutes 41 seconds

Keep a diary and one day it'll keep you.

Mae West

***

Proportion of men's clothes purchased by women today:
30 per cent
in 1985:
60 per cent

Proportion of grocery shoppers who replace an unwanted item back where they found it:
33 per cent

Proportion of people with tattoos who say they regret having them:
17 per cent

Longest time spent living in a tree (by Julia Hill, 1997–1999):
737 days
height off the ground:
48.8 metres

Proportion of surveyed women who nearly always sit with their legs crossed:
50 per cent

Proportion of newlyweds who do not have sex on their wedding night:
30 per cent

Number of words spoken each day by the average person:
about 7,500

Proportion of all MENSA (high IQ organisation) members who are left-handed:
20 per cent

Proportion of Britons who believe in psychic ability:
60 per cent
who have had a premonition of an event:
50 per cent

Average number of squares of toilet paper used by Americans each day: **57**

Proportion of the world's population who can roll their tongue:
64 per cent

Proportion of the world's population who will live to be 116 or older:
1 in 2 billion

The life span of a taste bud:
about 10 days

Number of cars that are added to the world's roads each year:
50 million

The times each day the average person laughs:
about 15

Average amount of water each person breathes and sweats out per hour on a plane:
100 millilitres

What he lacks in intelligence, he makes up for in stupidity.

Anon.

***

Average amount by which an unattended beard will grow each year:
14 centimetres

Proportion of the average fart that is nitrogen: **59 per cent**
hydrogen: **21 per cent**
carbon dioxide: **9 per cent**
propane: **7 per cent**
oxygen: **4 per cent**

Proportion of American women who say they rarely wear underwear:
4 per cent

Height of Beethoven:
165 centimetres

Number of men in the Bolivian navy in 2008: **about 5000**
length of their coastline: **0**

Proportion of Canadian women who would rather eat chocolate than have sex:
38 per cent

Number of coins found in the stomach of a 62-year-old man in France in 2004: **350**
total worth: **$A820**
total weight: **5.4 kilograms**

Number of letters Lewis Carroll (author of *Alice in Wonderland*) wrote to his friends in the final 37 years of his life:
98,721

Proportion of Americans who think the Sun revolves around the Earth:
20 per cent

Proportion of women in the USA with a bra cup size:
A: **15 per cent**
B: **44 per cent**
C: **28 per cent**
D: **10 per cent**
other: **3 per cent**

Number which sufferers of Triskaidekaphobia have a fear of:
13

Longest distance to spit a cherry stone (Horst Ortmann, Germany, 1994):
28.98 metres

Proportion of road accidents in Canada that involve a moose:
0.3 per cent

Proportion of people who say they never or rarely dream in colour:
18 per cent

Longest time standing motionless (Arulananthan Suresh Joachim):
21 hours 30 minutes

The secret to life is honesty and fair dealing. If you can fake that, you've got it made.

Groucho Marx

***

Proportion of Americans who would risk their lives to save their pet:
85 per cent

Estimated number of murders in 2008 per 100,000 population in:
Honduras: **58 (world's highest)**
South Africa: **37.3**
Mexico: **10**
Thailand: **8**
USA: **5.8**
France: **1.7**
Australia: **1.2**
UK: **5.8**
New Zealand: **2**
Japan: **0.5**
Russia: **2.3**
Sri Lanka: **6.7**
Singapore: **0.4**

Proportion of Americans who flush the toilet while sitting on it:
1 in 3

Proportion of women worldwide with naturally blonde hair:
2 per cent

Average length fingernails grow in a week:
0.5 millimetres

Weight of dead skin an adult sheds each year:
about 4 kilograms

Average length of a blink:
0.3 – 0.4 seconds

World record for continuously staying awake (Randy Gardner, a 17-year-old, USA, in 1964):
264 hours

Average duration of a yawn:
6 seconds

Weight of paper used in the 2006 Australian Census guides:
500 tonnes
date of the Census:
8th August 2006

Number of women in the USA in 2008 who had breast enlargements:
335,671

Proportion of Americans who believe in ESP:
60 per cent
in lucky numbers:
32 per cent
who think astrology is scientific:
40 per cent
accept alternative medicine:
88 per cent

World land speed record (Andy Green, UK, in 1997):
1,228 kilometres per hour

A girl phoned me the other day and said ... 'Come on over, there's nobody home'. I went over. Nobody was home.

Rodney Dangerfield

***

Probability that a woman will buy something she tries on in a store: **25 per cent**
a man: **60 per cent**

Proportion of American men who are happy with the size of their partner's breasts:
71 per cent

Proportion of the world's population who usually eat using chopsticks:
37 per cent

Proportion of women in Switzerland who secretly check their partner's mobile phone for evidence of an affair: **45 per cent**
of men who check: **31 per cent**

Proportion of surveyed Australians who believe that aliens have visited Earth:
67 per cent
that aliens exist:
80 per cent

that authorities are covering up evidence of aliens:
5 per cent

Proportion of US shoppers who turn right after entering a store:
85 per cent

Weight of the winner of 'Miss Jumbo Universe' in 2005
(Thanchanok Mekkeaw): **182 kilograms**
her age: **25 years**

Number of cups of coffee drunk worldwide each year:
over 400 billion

Time it can take the sting of a Box Jellyfish to kill a human:
3 minutes

Proportion of extra-marital affairs in Italy that are uncovered by evidence left on mobile phones:
87 per cent

Number of rat catchers employed in the Rat Surveillance Department in New Delhi: **97**
total number of rats caught by them in the last ten years: **0**

RATS.

My opinions may have changed, but not the fact that I am right.

Ashleigh Brilliant

***

Number of Australians to have won the men's US Open Championship golf tournament:
2 (David Graham in 1981, Geoff Ogilvy in 2006)

Speed limit of cars driven in the city in the UK in 1878:
3 kilometres per hour
in the country:
6 kilometres per hour

Estimated number of different gangs in the USA: **21,500**
number of members: **731,000**

Number of US Presidents to have obtained a patent:
1 (Abraham Lincoln in 1849 for a device to lift boats over shoals)
Patent Number: **6469**

Amount suggested Australian women be awarded for a three-minute lap dance: **$310**
a topless dance: **$510**
a fifteen-minute explicit tabletop dance: **$1,960**

working as a nude waitress for an hour: **$1,100**

Amount awarded in compensation to a Hong Kong construction worker who cannot stop swearing since being hit on the head by a plank:
$A1.95 million

Average weight of a US *Playboy Playmate:*
52 kilograms
the average American woman in 2009:
75 kilograms

Proportion of Australians who say they do not have a high sex drive:
10 per cent
Japanese:
13 per cent (highest)
Indians:
3 per cent (lowest)

Proportion of men in the USA who cried at their wedding:
25 per cent
of women:
45 per cent

Average age of Australians who had a facelift in 2006:
45 years
in 2003: **55 years**

Loudest recorded scream (by Jill Drake, London, 2000):
129 decibels
her age at the time:
48 years

Average reading speed: **200–250 words per minute**

Number of devices patented in the USA to prevent snoring:
over 300

If at first you don't succeed, try, try again. Then give up.

There's no use in being a damn fool about it.

W.C. Fields

***

Proportion of London Underground that is not underground:
58 per cent

Riskiest time for a car crash:
11pm to 6am
a pedestrian fatality:
6pm to 12 midnight

Average speed at which a bullet leaves the muzzle of a rifle:
3,000 kilometres per hour

Proportion of Australian men who wash their hands after using a public toilet: **75 per cent**
of women: **90 per cent**

Average number of breaths taken by a person living to age 80:
580 – 670 million

Average time a person spends on the toilet in a lifetime:
about 9,000 hours

Proportion of surveyed women who have hovered over public toilets rather than sitting on them:
80 per cent

Proportion of Americans who send Christmas cards to their cat or dog:
55 per cent

Number of Charles Dickens' novels to have gone out of print: **0**

Proportion of extramarital affairs that occur between noon and 3.00pm (mostly in hotels):
50 per cent

Average number of times a woman changes her hairstyle
between the ages of 18 and 24: **20**
between the ages of 50 and 80: **4**

Most hot chillies eaten in 3 minutes: **550**

Average number of times a woman falls in love in her lifetime:
5

Weight of the 23-year-old winner of Thailand's 'Miss Jumbo Queen' beauty contest in 2003 (the winner is the one who best exhibits the characteristics of an elephant, by virtue of her grace, elegance and size): **165 kilograms**
in 2005: **110 kilograms**

Everyone has the ability of making someone happy, some by entering the room, others by leaving it.

Anon.

***

Estimated volume of dog urine left in London's parks each year:
4 million litres

Average number of glasses of milk produced by a cow in her lifetime:
70,000

Number of people declaring themselves "witches" in the 2006 Australian Census:
8,210

Number of live, medium-sized cockroaches eaten in one minute (by Ken Edwards, England): **36**
year: **2001**

Most push-ups completed in one hour (Roy Berger, Canada, 1998):
3,416

Longest kiss (Germany, 2009, between Nikola Matovic and Kristina Reinhart):
32 hours 7 minutes 14 seconds

Number of crown-top bottles David Lambert opened with his teeth in one minute (Sweden, 2001):
50

Average number of peanuts required to make 1 kilogram of peanut butter:
about 1,600

Proportion of Americans who have signed their pet's name on a greeting card:
72 per cent

Most bricks balanced on a head for 10 seconds (John Evans, UK, 1998):
101
total brick weight:
188.7 kilograms

Smallest waist measurement of a living adult (Cathie Jung, USA):
38.1 centimetres

Estimated number of people worldwide who are without hygienic toilets:
2.6 billion
proportion of the world's population:
40 per cent

Length of time frozen sperm can survive:
between 10 and 15 years

I like work; it fascinates me. I can sit and look at it for hours.

Jerome K. Jerome

***

Proportion of dollar bills in the USA that contain traces of cocaine:
90 per cent
average time in circulation:
20 months

Ratio of sheep to people in New Zealand in 2009: **7 to 1**
in Australia: **4 to 1**
in the Falkland Islands: **160 to 1**

Average time at which people begin to get impatient when
waiting for a lift:
35 seconds

Sum of the numbers on a roulette wheel:
666

Average amount a person shrinks between the ages of 30 and 40 years:
1.5 centimetres

Number of psychiatrists per million people in the USA: **143**
in China: **15**

Proportion of women who say their favourite thing to do in bed is read:
75 per cent

Proportion of people in England who say they have personally experienced a ghost:
13 per cent

In a five-card combination dealt from a deck of 52 cards, the chance of getting a:
Royal flush: **1 in 649,740**
full house: **1 in 694**
straight: **1 in 255**
three of a kind: **1 in 47**
two pairs: **1 in 21**
one pair: **1 in 2**

Amount by which human brains have shrunk in the past 30,000 years:
10–15 per cent

Proportion of pregnancies worldwide that end in a live birth:
63 per cent
a miscarriage/stillbirth: **15 per cent**
induced abortion: **22 per cent**

Most likely time for engaged couples to break-up:
9.00pm to midnight

Average number of times a sleeper rolls over each night:
12

Time at which turtles mate:
about 4.00pm

What's the three words you never want to hear while making love?

Honey, I'm home.

Ken Hammond

***

Number of 100 clay targets shot by US cowgirl Annie Oakley (1860–1926) from a distance of 14.5 metres in 1922:
100 (a record)

Number of men gunned down in the 1929 St Valentine's Day Massacre: **7**

Number of different living languages in the world:
about 6,500

Odds of the racehorse Ajax chasing his 19th successive victory when he lost the Rawson Stakes at Rosehill, New South Wales:
40–1 on
the year: **1939**
number of horses in the race: **3**

Average number of hairs in an eyebrow:
450–550

Time since the last Ice Age on Earth:

10,000–20,000 years

Total number of airline passengers on flights in the world in 2006: **2.13 billion**
number of bags checked in: **300 million**
number that go astray: **30 million**
number that are never recovered: **200,000**
forecast number of airline passengers on flights in 2011:
2.75 billion

Altitude of the world's highest airport (Bangda, Tibet):
4,334 metres

Average amount of saliva a person produces in a lifetime:
about 38,000 litres

Age of legendary Marshall Wyatt Earp when he died in his sleep in 1929:
80 years

Proportion of muscle that men lose between the ages of 30 and 80 years:
40 per cent

Age of the oldest person to be elected President of the USA (Ronald Reagan, 1980): **69 years**
oldest person to hold office (Ronald Reagan, 1988):
77 years

Reported weight of Albert Einstein’s brain: **1.49 kilograms**
year he died: **1955**
his age: **76**

Flight by machines heavier than air is unpractical and insignificant, if not utterly impossible.

Simon Newcomb,* Professor of Mathematics and Astronomy, *1902

***

Average number of times per minute you blink when relaxed:
15
squinting: **6**

Average time to say a sentence during normal conversation:
2–3 seconds

The number between 18 and 20 on a dartboard: **1**

Proportion of Americans who would like to have more novelty in their lives:
61 per cent
who are bored with television and the internet:
80 per cent

Estimated volume of gold ever mined:
155,000 tonnes
proportion since 1950:
65 per cent

Wingspan of a Boeing 747–400 jumbo jet:
65 metres
empty weight:
320 tonnes

Proportion of the world's precious opals produced in Australia:
95 per cent
black opals:
99 per cent

Overall length of the *Oasis of the Seas* (the world's largest cruise liner): **360 metres**
height above water line: **72 metres**
weight: **225,000 tonnes**
number of decks: **16**

Number of bees that broke free when a lorry carrying hives crashed in Florida in 2003:
80 million

Number of moustachioed kings in a standard deck of playing cards:
3 (the King of Hearts does not have one)

Longest fall survived in a lift (Empire State Building, New York):
75 storeys
distance: **300 metres**
year: **1945**

Longest time spent balancing on one foot (Arulanantham Suresh Joachim, 1997):
76 hours 40 minutes

Number of British stamps that carry the name of their country:
0

Listen son, you ain't goin' nowhere. You oughta go back to drivin' a truck.

Jim Denny,* Manager of the Grand Ole Opry, after Elvis Presley's first performance, *1954

***

Point at which sound can cause permanent damage to hearing:
85 decibels

Number of curls in the young actor Shirley Temple's hair: **56**

Number of peas in a pod:
7–9

Proportion of women who have a tattoo on a breast that have it on their left breast:
75 per cent

Proportion of people who use toilet paper for blowing or wiping their nose:
60 per cent

Average height of US Playboy Playmates: **167.6 centimetres**
measurements: **35–23–35 (inches)**
proportion who are blonde: **40 per cent**

oldest: **60 years (Patricia Paay, December 2009)**
youngest: **17 years (Elizabeth Ann Roberts, January 1958)**
average age when they pose: **22 years**

Average length of a wire coathanger when straightened:
112 centimetres

Tilt required on a bowling pin for it to tip over:
7.5 degrees

Average weight of the ashes of a cremated person:
4 kilograms

Average weight of a medieval knight's suit of armour:
24 kilograms

Speed at which a sneeze leaves the nose:
about 160 kilometres per hour

Number of cards in a full tarot pack: **78**

Number of rubber ducks believed to be floating on the world's oceans: **29,000**
number of Lego pieces: **5 million**

!
DESERTED ISLAND

The motor car will never usurp the place of the horse.

The Economist, 1911

***

Time it took bantamweight boxer Jimmy Carruthers to win Australia's first boxing world title (against Vic Toweel of South Africa): **2 minutes 19 seconds**
his age: **23 years**
number of landed punches thrown by Carruthers: **147**
number of landed punches thrown by Toweel: **0**

Number of digits on each hand of Mickey Mouse: **4**
of Homer Simpson: **4**

Greatest height from which a fresh egg has been dropped and remained intact:
213 metres

Average speed of a crack in breaking glass:
5,000 kilometres per hour

Amount paid in 2004 for the 'lucky' phone number 13585858585 on a Chinese auction site:
$A1.5 million

Amount for which a woman sued a hotel chain after she developed a "toilet phobia" after falling off one in a South African hotel in 2004:

$A20,000

RELATIONSHIPS

Human relationships would be a lot better if people weren't involved.

***Lindsay Foyle,* The Australian**

He's turned life around. He used to be depressed and miserable.

Now he's miserable and depressed.

David Frost,* TVAM, *1984

***

Proportion of women who feel they look natural:
31 per cent
average: **29 per cent**
beautiful: **2 per cent**

Proportion of overseas romances that are over within a week of returning home:
90 per cent

Proportion of women who consider themselves their most beautiful in their:
twenties: **28 per cent**
thirties: **50 per cent**
forties: **16 per cent**
other: **6 per cent**

Proportion of men who feel that blondes look friendly and approachable:
63 per cent
are needy and lacking in independence:
40 per cent

Proportion of women who think that men aren't really interested in finding out what is wrong with them when they are upset:
66 per cent

Proportion of surveyed men who say they prefer fake large breasts to smaller real ones:
22 per cent

Proportion of women who say they are secretly flattered at the thought of men fighting over them:
65 per cent

Proportion of surveyed British people who would like to be rid of their current partner:
8 per cent

Proportion of men who say that the feature that attracts them most about women is their:
breasts: **17 per cent**
backside: **17 per cent**
face: **14 per cent**
personality: **11 per cent**

Proportion of surveyed 18–35-year-olds in Australia who have dumped a partner via a text message:
15 per cent

Estimated proportion of people who use personal ads for dating who are already married:
20–35 per cent

Proportion of women who don't expect a man to ask them out on a second date if they have sex on the first date:
80 per cent

Thou shall not commit adultery ... unless in the mood.

W.C. Fields

***

Proportion of women who keep purchases secret from their partners:
32 per cent

Proportion of surveyed married or cohabiting men without academic qualifications who had an affair in the previous five years: **12 per cent**
without tertiary qualifications: **6 per cent**

Proportion of single males in the USA who met someone they were interested in but did not act on it: **75 per cent**
of married males: **58 per cent**

Proportion of Australian men aged 16–60 who have cheated on their partner in the past year: **5 per cent**
women: **3 per cent**

Average age at which people worldwide have sex for the first time:
17 years 4 months

Proportion of men who think a woman's invitation for coffee is really an invitation to have sex:
68 per cent

Proportion of women who say they flirt at work:
80 per cent
who say that a male colleague has made a pass at them:
80 per cent

Proportion of men who say that 'fun-loving' women attract them the most:
67 per cent
proportion who say they are most attracted to assertive women:
6 per cent

Proportion of women in Australia marrying younger men:
20 per cent
in 1974: **11 per cent**

Proportion of surveyed Australians who said they would like to spend more time with their:
romantic partner: **56 per cent**
friends: **27 per cent**
family: **12 per cent**
children: **4 per cent**

Proportion of women who decide whether they want to see a man again within 10 minutes of the first date:
1 in 3

Reported number of ways to keep an adulterous affair secret:
91

Your friend is the man who knows all about you, and still likes you.

Elbert Hubbard,* The Notebook, *1927

***

Proportion of Australian couples who say the most negative influence on their relationship is:
lack of time: **38 per cent**
lack of money: **20 per cent**
children: **14 per cent**

Proportion of Australians who feel that lack of compromise between men and women both wanting careers puts relationships in jeopardy:
90 per cent

Rise in testosterone levels of males aged 18–36 after a five-minute chat with a young female:
30 per cent

Proportion of Australian males with erectile problems:
nearly 40 per cent

Proportion of young Australian couples who talk to their partner about sex at least once a week:
63 per cent

Estimated proportion of people in Australia with herpes:
17 per cent

Proportion of surveyed men in the USA who say that the most desirable thing in a romance is:
'emotional intimacy': **31 per cent**
'sex': **23 per cent**

Proportion of separated people in Australia who say they feel unhappy or terrible: **10 per cent**
of married people: **2 per cent**

Proportion of Australian couples who rate companionship and friendship (the highest factor) as the best thing about their relationship: **28 per cent**
being in love: **5 per cent**

Proportion of women who say they would confess an affair with another woman to their male partner:
50 per cent

Proportion of Australians who say they want to spend more time with their families:
75 per cent

Proportion of surveyed Britons who have dumped a boyfriend/girlfriend just before Christmas or Valentines Day just to avoid buying a gift:
5 per cent

Laugh and the world laughs with you, snore and you snore alone.

Anthony Burgess,* Inside Mr Enderby, *1968

***

Proportion of surveyed Australian men who have had an affair with a work colleague:
60 per cent

Proportion of office relationships that survive more than six months:
1 per cent

Proportion of men who feel they can steer a shopping trolley better than a woman can:
70 per cent

Proportion of women who feel they can steer a shopping trolley better than a man can:
60 per cent

Proportion of women who say they would feel less guilty about cheating with another female than with a man:
90 per cent

Proportion of Australians who have had romantic moments interrupted by heartburn:

10 per cent

Proportion of Australians aged over 75 who believe they are 'middle-aged':
20 per cent
who consider themselves very attractive for their age:
30 per cent

Proportion of workers who say they would cover up for their boss if it meant keeping their job:
30 per cent

Proportion of men in the US who say their mother favoured them: **66 per cent**
of women: **27 per cent**

Proportion of British men who say they cry over love songs:
25 per cent
of women: **33 per cent**

Proportion of British dog owners who would sometimes rather spend time with their pet than other people:
46 per cent

Proportion of surveyed adults who have had a one-night stand:
Australians: **60 per cent**
Norwegians: **70 per cent (highest)**
Indians: **13 per cent (lowest)**

It is only shallow people who do not judge by appearances.

***Oscar Wilde,* The Picture of Dorian Gray**

***

Proportion of British women who say they tell white lies:
to their partner: **70 per cent**
to their friends: **67 per cent**

Proportion of men who like the way they look naked: **68 per cent**
of women: **22 per cent**

Proportion of unfaithful husbands who don't leave their wives:
85 per cent

Proportion of surveyed adults who have had sex in a car:
Australians: **73 per cent**
Italians: **82 per cent (highest)**
Chinese: **12 per cent (lowest)**

Proportion of women who say that love is more important to them than sex, power or wealth: **89 per cent**
of men: **65 per cent**

Proportion of working people who report having had an office affair:
30 per cent
proportion of these affairs that ended with one of the parties leaving the organisation:
73 per cent

Proportion of Australian women who say that work stops partners spending more time with them:
60 per cent

If a man and a woman fight on the telephone the proportion of time that it will be the woman who hangs up first:
83 per cent

Proportion of surveyed British children who feel that their mother prefers the cat to their father:
33 per cent

Proportion of Americans who have ended a romantic relationship because of a cat:
11 per cent

Proportion of Australians who say they are lonely: **33 per cent**
age at which they are the loneliest: **40s**

Proportion of Australian males who say their job has a bad effect on their sex life:

43 per cent
of Australian females:
47 per cent
age group who suffer the most:
25–54 years

To love oneself is the beginning of a lifelong romance.

Oscar Wilde

***

Proportion of women who would rather not know the true reason their partner ended the relationship:
33 per cent

Proportion of Australians who describe themselves as 'shy':
30–40 per cent
proportion who actually behave in a shy manner:
15–20 per cent

Proportion of Australian men who believe unhappy couples should stay together for the sake of the children: **24 per cent**
women: **15 per cent**

Proportion of Australian men who refuse to pay for their partner's dinner on a first date:
25 per cent
pay only after their partner offered:
30 per cent
always pay:
45 per cent

Proportion of Australian women who say they sleep in the nude with their lovers: **50 per cent**
when alone: **33 per cent**

Proportion of British women who when they wake up:
want a cup of tea: **52 per cent**
want sex: **3 per cent**

Proportion of women who say they indulge in retail therapy when in a bad mood:
40 per cent

Proportion of married men in Australia who will outlive their wives:
22 per cent

Number of 'lords-a-leaping my true love gave to me':
10

Number of people in Hong Kong who have had just one sexual partner:
65 per cent (world's highest)

How much more likely it is to strike up a conversation or a relationship when you walk in the park with a dog rather than alone: **10 times**

BABIES

A baby will make love stronger, days shorter, nights longer, bankrolls smaller, a home happier, clothes shabbier, the past forgotten and the future worth living for.

Anonymous

Having one child makes you a parent; having two you are a referee.

David Frost,* Independent, 16 September *1989

***

Age at which a foetus develops some basic breast tissue:
6 weeks

Number of times each day a newborn baby urinates:
up to 18

Fertility rate per female in Australia in 2009: **1.78**
in 1960: **3.50**
level required to replace the population: **2.10**

Proportion of three-year-olds who still wet the bed at least once a week:
25 per cent
five-year-olds:
10 per cent

Average weight of a single newborn baby:
3.35 kilograms
average weight of a newborn triplet:
1.70 kilograms

Average amount by which a newborn baby increases in length in the first month:
3.5 centimetres

Proportion of Australian mothers who breastfeed their babies beyond six months:
20 per cent
Proportion of babies in Australia that are conceived through IVF:
3 per cent

Proportion of babies affected by infantile colic in their first six months:
28 per cent
age at which symptoms usually peak:
6 weeks

Age at which nappy rash occurs most often:
after 8 months

Number of babies who died in Australia in 2007 from SIDS: **71**
in 1991: **131**

Proportion by which a baby's weight decreases immediately after birth:
10 per cent
age at which it returns to birth weight:
2 weeks

Average sleep each day of a newborn baby:
16 hours

Distance from their face at which a newborn baby can see objects most clearly:
20–30 centimetres

If I were younger, I'd know more.

James Barrie

***

Most likely age group for a woman in Australia to have a baby:
30–34 years
babies per 1000 women in this age group:
111 babies

Number of bones with which babies are born: **300**
number they have by adulthood (some bones fuse together):
206

Proportion of a newborn baby's body weight accounted for by its head:
25 per cent

The largest number of babies born to one woman (a Russian woman, 1725–1765): **69**
number of sets of twins she had: **16**
sets of triplets: **7**
sets of quadruplets: **4**

Proportion of full-term babies that have jaundice:
50 per cent

Average beats per minute of a foetal heart:
120 to 160

Amount spent in Australia each year on:
disposable nappies: **about $415 million**
baby formula: **$140 million**
prams: **$90 million**
cots: **$70 million**
high chairs: **$35 million**

Proportion of new mothers who suffer from postpartum
depression: **about 50 per cent**
day at which it typically occurs following delivery: **3**

Age at which children can be taught to use a toothbrush:
2 years

Proportion of Australian women currently in their childbearing years who will never have children:
25 per cent

Number of babies a female mouse can have in a year:
up to 100

Proportion of babies in Australia who are breastfed on being discharged from hospital: **83 per cent**
at 26 weeks: **49 per cent**

Proportion of 0–3-year-olds that were breastfed at some time:
87 per cent

People who say they sleep like babies usually don't have them

Leo J. Burke

***

Proportion of babies aged under three months that are being fully breastfed:
64 per cent

Number of women in the world of child-bearing age (15–44 years):
1.4 billion

Number of cells in a newborn baby: **26 billion**
in an adult: **50–100 trillion**

Babies per 1000 women aged over 40 in Australia in 2006: **2.6**
in 1982: **1.3**

Combined weight of the world's lightest surviving triplets (named Coffey, born 1998, USA):
1,385 grams

Age of the oldest woman (Omkari Panwar of India) to give birth (to twins): **70**
year: **2008**

Number of times per day that babies laugh spontaneously:
up to 300
by adulthood: **15**

Estimated proportion of male Australian babies who are circumcised:
10 per cent

Proportion of babies born to Australian women aged over 35:
18.5 per cent
in 2000: **15.7 per cent**

Average amount of time a newborn baby cries each day:
113 minutes

Proportion of surveyed women in Scotland who went on to have another baby after delivering their previous baby by caesarean section:
67 per cent
after a natural delivery:
72 per cent

Estimated number of unplanned pregnancies worldwide each year:
180 million

Time that a baby's fingernails have been growing when born:
15 weeks

Age at which most babies can recognise their own name:
4–7 months

Average age of midwives in Australia: **41 years**
reported shortage: **1,850**

First you have to teach a child to talk, then you have to teach it to be quiet.

Herbert Prochnow

***

Probability of a child being left-handed if:
both parents are right-handed: **9.5 per cent**
one parent is left-handed: **19.5 per cent**
both parents are left-handed: **26 per cent**

Estimated proportion of all humans ever born who are alive now:
10 per cent

Number of treatment embryos reportedly in storage in Australia and New Zealand: **about 105,000**
in 1994: **about 22,000**

Age at which babies show a preference for their mother's face:
2 weeks

Proportion of a baby's brain nerve cell connections that are made in the two weeks after birth:
90 per cent

Age at which most babies begin smiling in a social way:

4–8 weeks

SPORTS

I'll never forget my first fight ... all of a sudden I found someone I knew in the fourth row. It was me.

Henny Youngman, 1940

Sure, there have been deaths and injuries in boxing, but none of them serious.

Alan Winter

***

Diameter of a tennis ball:
6.35 centimetres
weight:
56.7 grams
number used at Wimbledon 2009 (including qualifying and practice matches):
54,200

Longest playing time in a cricket Test match (South Africa v England, Durban, 1939):
43 hours 16 minutes
total runs scored:
1,981

Estimated number of soccer players in the world:
about 250 million
number of countries (membership of FIFA):
208

Height of a tennis net at the centre: **0.914 metres**
at the posts: **1.07 metres**

In baseball, distance from the home plate to the pitching
mound: **18.45 metres**
from home plate to second base: **38.8 metres**
from base to base: **27.43 metres**

Width of a badminton singles court: **5.18 metres**
length: **13.40 metres**

Highest speed reached in a speed skydiving competition (by France's Michael Brooke, 2002):
524.13 kilometres per hour

Most ducks in Test cricket (by Courtney Walsh, West Indies): **43**
number of Tests: **132**
number of innings: **185**
his highest score: **30 not out**
his batting average: **7.54**

Longest time to spin a regular basketball on one finger (Zhao Guang, China, 2003):
3 hours 59 minutes

Number of Tests played by Australian cricketer Dennis Lillee: **70**
wickets taken: **355**
bowling average: **23.92**

Number of nations to have had a men's Wimbledon singles tennis champion: **13**

Largest field in a Melbourne Cup (in 1890): **39**
smallest field (in 1863): **7**

The minute you start talking about what you're going to do if you lose, you have lost.

George Shultz

***

Grams of gold in an Olympic gold medal: **6**

Maximum permitted length of a baseball bat:
1,067 millimetres

Proportion of British men who have taken their football team's shirt with them on holidays:
14 per cent

Largest victory in a major golf tournament (by Tiger Woods,
2000 US Open): **15 shots**
his score: **272**
amount under par: **12**

Diameter of a mens' discus: **21.8 centimetres**
womens' discus: **18.3 centimetres**

Most Olympic medals won by an Australian (swimmer Ian Thorpe): **9**
number that were gold medals: **5**
silver: **3**
bronze: **1**

Number of tricks for a small slam in bridge: **12**
a grand slam: **13**

Number of Olympic swimming medals won by Michael Phelps (USA) as at 2010: **16**
number that were gold: **14**

Height of a cricket wicket above the playing surface: **71.12 centimetres**
width: **22.86 centimetres**

Number of world billiard records held by Australian Walter Lindrum when he retired undefeated in 1950: **57**

Number of first-class centuries scored by cricketer Don Bradman: **117**
number of these that were double centuries: **37**
that were Test centuries: **29**

Lowest completed team innings score in Test cricket (by New Zealand against England in 1954–55): **26**

Largest margin of victory in a one-day cricket international (New Zealand v Ireland in Aberdeen, 2008): **290 runs**
New Zealand's score (for two wickets): **402**
Ireland's score (all out): **112**

I went to a fight the other night and a hockey game broke out.

***Rodney Dangerfield,* comedian**

***

Total prize money for the Australian Tennis Open
in 2010:
$21,400,000
in 1976: **$85,000**

Number of stitches on a baseball: **108**
The number of laps to complete the Indianapolis
500 auto
race: **200**

Diameter of a golf hole:
10.8 centimetres

Number of Major League seasons played by
baseballer
Babe Ruth: **22**
batting average: **0.314**
hits: **2873**
home runs: **714**
times struck out: **2,213**

Number of dimples generally on a golf ball:
336 (US) or 330 (UK)

Number that lies between 20 and 12 on a dartboard: **5**

Number of consecutive 400 metre hurdle races won by Edwin Moses, USA, between 1977 and 1987:
122

Number on David Beckham's shirt for Spanish soccer club
Real Madrid: **23**

World record for 100 metres running for men aged 95–99 (Kozo Haraguchi, Japan, 2005):
22.04 seconds

Greatest distance skated on a longboard in 24 hours (Paul Kent, Seattle, 2009):
403 kilometres

Year of the cricket 'underarm' bowling incident:
1981

Height of the men's high hurdles:
1.067 centimetres
distance to the first hurdle in men's 110 metres race:
13.72 metres
between the next 9 hurdles:
9.14 metres

Number of competitors at the Commonwealth Games in 2006:
around 4,500
in 2002: **4,250**
in 1930: **40**

Most points scored in a National Rugby League (NRL) match in Australia:
102 (Newcastle Knights 70 def Canberra Raiders 32)
season: **2006**

Some people think football is a matter of life and death. I assure you, it's much more serious than that.

Bill Shankly

***

Lowest weight ever carried by a horse to win a Melbourne Cup (Banker in 1863):
33.5 kilograms
highest weight (Carbine in 1890):
66.0 kilograms

Number of successive Olympics at which the Hungarian sabre fencer, Aladar Gerevich, won gold medals (between 1932 and 1960): **6**

Time it takes a baseball pitched at 150 kilometres per hour to reach home plate:
0.4 seconds

Number of times Australia has won the Rugby Union World Cup:
2 (1991 and 1999)
Rugby League World Cup:
9 (1957, 1968, 1970, 1975, 1988, 1992, 1995, 2000)
Cricket World Cup:
4 (1987, 1999, 2003, 2007)
Soccer World Cup: **0**

Number of athletes to win a gold medal at both the Summer and Winter Olympics:
1 (Eddie Eagan, USA, boxing in 1920 and bobsled in 1932)

Most individual medals won at the Olympic Games (by gymnast Larissa Latynia, USSR): **18**
number that were gold medals: **9**
years: **1956–1964**

World record for a one mile run in:
1915: **4 minutes 12.5 seconds**
1935: **4 minutes 6.4 seconds**
1955: **3 minutes 57.9 seconds**
1975: **3 minutes 49.4 seconds**
1995: **3 minutes 44.9 seconds**
2005: **3 minutes 43.13 seconds (set in 1999 and still the record)**

Number of sets it took Australian Pat Cash to beat Czech player Ivan Lendl to win the 1987 men's Wimbledon singles tennis title:
3 (7–6, 6–2, 7–5)

Fastest 100-metre sprint on a unicycle (Peter Rosendahl, USA):
12.11 seconds

Highest Test cricket score by a team (Sri Lanka v India, 1997, loss of six wickets):

952

Minimum number of balls in a croquet set: **4**

Proportion of skiing accidents in Switzerland that involve an injury to the knee:
35 per cent

Serious sport has nothing to do with fair play. It is bound up with hatred, jealousy, boastfulness and disregard for all the rules.

George Orwell, 1950

***

Average speed at which a dart hits a dartboard during competition play:
65 kilometres per hour

Number of kilojoules burned when driving a car for one hour: **0.50**
playing tennis for one hour: **about 500**

Number of Americans who play soccer:
about 18 million
number of registered players worldwide:
about 260 million

Distance of the Olympic marathon:
42,195 metres

Year the first man (Johnny Weissmuller) swam 100 metres in under one minute: **1922**
his time: **58.6 seconds**
current world record (César Cielo, Italy, 2009): **46.91 seconds**

distance by which Cielo would have beaten Weissmuller over 100 metres based on these times: **about 20 metres**

Age of English cricketer Jack Hobbs when he scored his 100th first class century: **40 years**
year: **1923**

Winning time of the men's 4x200 metre freestyle swimming relay at the 1924 Paris Olympics (a then world record):
9 minutes 53.4 seconds
current men's world record (in 2010):
6 minutes 58.55 seconds
current women's world record (in 2010):
7 minutes 42.08 seconds

Number of runs that 17-year-old Don Bradman scored for Bowral against Moss Vale in the Berrima District Final:
300
year: **1926**

Number of gold medals won by athlete Jesse Owens at the 1936 Berlin Olympics:
4 (100 metres, 200 metres, 4x100 metre relay, long jump)
length of time his long jump stood as a world record:
25 years

Size of the crowd at Caulfield Racecourse Melbourne in October 1946 to see Bernborough run in the Caulfield Cup:
108,213
weight carried: **68 kilograms**
finishing position: **5**

There's been a heavy fall of rain here at Trent Bridge but fortunately it didn't touch the ground.

***Brian Johnston,* English cricket commentator**

***

Year that polo was last played as an Olympic sport (won by Argentina):
1936

Don Bradman's score in his last innings in Test Cricket, played at The Oval, England in 1948: **0**
his last innings in first class cricket (South Australia v Victoria, 1949): **30**

Age of Ken Rosewall when he won the Australian men's singles tennis title (as the youngest winner):
18 years

Number of times that Australian golfer Peter Thomson won the British Open: **5**
year of his first win: **1954**
last win: **1965**

Age of the youngest Formula One champion (Lewis Hamilton, Britain, in 2008):
23 years, 301 days

Score by which Australia defeated the British Isles at the Sydney Cricket Ground in the First Rugby Union Test in 1899:
13–3

Diameter of a basketball ring:
45.72 centimetres
height from the ground:
3.05 metres: (10 feet)

Proportion of British football fans who believe that club crowds have some influence over the decisions of the referee:
69 per cent

Year that a sport first had a world championship (billiards):
1873

Number of players in a field hockey team: **6**
ice hockey: **11**

Highest possible sore with three darts:
180

Number of balls in snooker: **22**
number of these that are red: **15**
total value of the coloured balls: **27**

Length of a cricket pitch:

20.12 metres (22 yards)
width:
3.05 metres (10 feet)

BIRDS, INSECTS, FISH AND OTHER ANIMALS

I have a memory like an elephant. In fact, elephants often consult me.

Noel Coward (attrib.)

Dogs come when they're called; cats take a message and get back to you.

Mary Bly

***

Number of eyes on a bee: **5**

Proportion of animal species that are insects: **75 per cent**

Length of the smallest full-grown dog in the world (a Chihuahua named 'Star'): **14 centimetres**
height: **7 centimetres**
weight: **454 grams**

Average brain weight of:
a sperm whale: **8.0 kilograms**
an elephant: **6.0 kilograms**
a human: **1.4 kilograms**
a bottlenose dolphin: **1.5 kilograms**
horse: **532 grams**
sheep: **140 grams**
cat: **30 grams**
rat: **2 grams**

Number of sheep and lambs in Australia in 2009–10 (the lowest since 1947):
about 70 million

Diameter of the eye of a giant squid:
40 centimetres

Average number of quills on a porcupine:
30,000

Average weight of the red kangaroo: **90 kilograms**
height: **1.85 metres**

Width of a bald eagle's nest: **90–150 centimetres**
depth: **90–185 centimetres**

Volume of water that an elephant can drink in one sitting:
up to 200 litres

Top speed of a rabbit:
72 kilometres per hour

I think animal testing is a terrible idea; they get all nervous and give the wrong answers.

Joseph Blosephina

***

Number of species of shark: **700**
number found in Australia: **160**

Rate at which a woodpecker can peck:
20–25 per second

Length of time an iguana can stay underwater:
28 minutes

Length of a fully-grown bottle-nose dolphin:
2.5–4.0 metres
weight: **up to 600 kilograms**
weight at birth: **about 10 kilograms**

Average life expectancy of a potbellied pig: **15 years**
height: **33 – 66 centimetres**

Number of eggs a turbot fish lays in its lifetime:
14 million

Estimated number of dogs in the USA that are currently mentioned in their owner's will:

1 million

Proportion of all road accidents in Sweden that involve an elk:
20 per cent

Number of different species of mosquito:
about 2,700

Average time a cockroach can live without a head:
7 days

Number of legs on a crab: **10**
on a lobster: **10**

Estimated number of tigers kept as pets in the USA:
5,000 – 7,000

Wingspan of the world's largest moth in overall size (Atlas moth):
30 centimetres

Highest recorded flight of a bird (Rupell's vulture):
11.3 kilometres

Proportion of the world's goats that live in India:
19 per cent
world's buffaloes:
54 per cent

Number of decks on Noah’s Ark: **3**

Number of kangaroos on an Australian $1 coin: **5**

I wonder if other dogs think poodles are members of a weird religious cult.

Rita Rudner

***

Average length of a black flying fox: **25 centimetres**
weight: **500–700 grams**
wingspan: **over 1 metre**

Sound level of the world's loudest insect (African cicada) at a distance of 50 centimetres:
106.7 decibels

Height of an adult giraffe (world's tallest mammal):
about 6 metres
at birth: **1.7–2.0 metres**

Average gestation period of the Asian elephant:
650 days

Longest ears measured on a rabbit:
79 centimetres

Average number of times each day a cat returns to its bowl:
36

Number of noses on a slug: **4**

Average life expectancy of a dog:
12 years

Weight of the world's heaviest insect (goliath beetle):
70–100 grams
length: **up to 11 centimetres**

Number of mammals that cannot jump:
1 (the elephant)

Number of species of frog:
over 4,000

Number of months an oyster can survive out of water:
4

Amount left in a will to a cat (named Rex) by his bachelor owner who died in Canada in 2005:
$A1.5 million

Number of days an ant can survive underwater:
2–14

Greatest recorded age of a snake (Popeye, a boa constrictor):
40 years 3 months 14 days

Number of knees on an elephant: **4**

Number of times the word 'dog' appears in the Bible:
18

Number of ants the South American giant anteater can eat in one day:
up to 30,000

I worked in a pet store and people would ask how big I would get.

Rodney Dangerfield

***

Average hours a koala sleeps each day: **22**

Deepest recorded dive of a turtle (a Leatherback): **1,211 metres**

Average wing flaps per second of a bee: **200–250**

Maximum voltage of an electric eel: **650 volts**

Number of teeth a blue whale has: **0**

Weight of an adult male ostrich (the world's largest living bird): **up to 160 kilograms**

Longest recorded length of a snake (*Python reticulatus* in 1912): **10 metres**

Year the cane toad was introduced to Queensland (to eat the Brayback beetle):

1935

Number of legs on a scorpion: **8**

Number of species of native bees in Australia: **about 1500**
number of these that do not sting: **about 10**

Equivalent human intelligence level of the average dog:
2 years old

Average number of words an intelligent dog can understand:
about 200

Number of species of earthworm:
over 3,000

The average life expectancy in captivity of: a beaver: **5 years**
a kangaroo: **7 years**
a leopard: **12 years**
a rhinoceros: **15 years**

Average speed of:
a garden snail: **0.05 kilometres per hour**
a three-toed sloth: **0.25 kilometres per hour** giant tortoise: **0.27 kilometres per hour**

Amount of venom from the Brazilian Wandering spider required to kill a human:
0.0005 grams

Number of ducks sold in China each year:
about 1.5 billion

Cats are smarter than dogs. You can't get eight cats to pull a sled through the snow.

Jeff Valdez

***

The longest recorded life span of a slug:
18 months

Average body temperature of a dog:
38–40°C

Number of pads on the front feet of a cat: **5**
back feet: **4**

Number of different types of bat: **over 900**
number who cannot fly: **0**

Fastest dive of a bird (a peregrine falcon):
up to 350 kilometres per hour

Average life span of an ostrich: **about 75 years**
age up to which it can reproduce: **50 years**

Number of teeth on an aardvark: **20**

Average weight of bamboo shoots eaten each day by a giant panda:
40 kilograms

Typical length of a tiger's tail:
90–120 centimetres

Number of pet goldfish in the UK:
15 million

Average number of trees a beaver can cut down in a year:
200

Average time it takes a lobster to grow to be 500 grams:
7 years

Proportion of a jellyfish that is water:
95 per cent

Most feathers ever counted on one bird (the Tundra Swan *Cygnus Columbianus*):
25,216
proportion that were on its head:
80 per cent

Estimated number of chickens eaten by humans worldwide every year:
10 billion

Length to which Burmese pythons can grow: **6 metres**
weight: **91 kilograms**
life expectancy: **25 years**

Proportion of the world's pigs that are kept by farmers in China:
50 per cent

God loved the birds and invented trees.
Man loved the birds and invented cages.

Jacques Deval

***

Average life span of a penguin:
6–7 years
oldest recorded age:
20 years

Number of species of piranha: **about 40**
number of these that are aggressive: **3**

Amount of milk each day a female whale can produce:
up to 600 litres

Number of stomachs a cow has: **4**

Top speed of a swordfish:
more than 105 kilometres per hour
maximum length:
4.6 metres
maximum weight:
650 kilograms

Number of species of ant in the world:
about 8,000

Average lifespan of a giraffe: **10 years**
a hippopotamus: **41 years**
a fox: **7 years**

Circumference of the world's largest wasp nest (Japan, 1999):
2.45 metres

Average number of honey bees in a hive:
about 40,000

Number of the world's top 15 venomous snakes that live in Australia: **11**

Average number of whiskers on a cat: **24**

Estimated number of humpback whales in the world:
about 20,000

Average reported deaths in Australia each year caused by horses: **20**
bees: **10**
snakes: **1.6**

Man is a clever animal who behaves like an imbecile.

Albert Schweitzer

***

Life expectancy of a black rhinoceros:
35–50 years
weight:
800–1,400 kilograms
length of horn:
up to 1.4 metres

Thickness of a hippopotamus skin:
up to 5 centimetres

Year the world's first monkey-crossing across a busy road was built (in Taiwan):
1998

Life expectancy of a moose:
15–25 years
maximum running speed:
55 kilometres per hour
swimming speed:
10 kilometres per hour

Average number of people bitten each day by animals (mainly dogs, horses, camels and donkeys) in Karachi, Pakistan:
120

Number of references to lions in the Bible: **176**

Proportion of polar bears that are left-handed:
100 per cent

Proportion of time that a walrus uses their right fin for eating rather than their left:
89 per cent

Estimated number of penguins in Antarctica:
20,000

World's buffalo population: **150 million**
proportion that are in India: **53 per cent**

Number of muscles in an elephant trunk: **40,000**
bones: **0**

Longest recorded lifespan of a goldfish:
41 years

Proportion of the world's insect species that are beetles:
27 per cent

Gestation period of a domestic dog: **61 days**
domestic cat: **63 days**

Number of animals in Antwerp Zoo in Belgium: **6,000**
number of different species (world's largest): **1,160**
year it was founded: **1843**

Fishing, with me, has always been an excuse to drink in the daytime.

Jimmy Cannon

***

Distance at which a slug can smell a mushroom: **2 metres**

Estimated number of spiders in Britain: **200,000 billion**

Number of animals that survived the journey of the First Fleet
in 1788:
geese **29**
ducks: **35**
fowls: **122**
chickens: **87**
stallions: **2**
sheep: **44**
goats: **19**
rats: **?**

Estimated number of species of fish: **23,000**

Number of eggs a female mackerel lays at one time:

400,000–500,000
their diameter:
0.97–1.38 millimetres

Length of a newborn kangaroo at birth:
about 2.5 centimetres

Time that a whale can go without eating:
up to 3 months

Amount of earth a single mole can tunnel through in a day:
20 metres

Number of dust mites on a typical bed mattress:
100,000–10,000,000

Average length of the erect penis of:
an adult gorilla: **5 centimetres**
a chimpanzee: **7.5 centimetres**

Average heartbeat per minute of a hedgehog:
190
during hibernation: **20**

Average weight of a beaver: **up to 27 kilograms**
length: **1.22 metres**

Average range of a pig squeal:
100–115 decibels

Length (nose to tail) of the African Gambian Giant Pouch rat (largest in the world): **up to 75 centimetres**
weight: **2 kilograms**
proportion that is tail: **50 per cent**

I am fond of pigs. Dogs look up to us. Cats look down on us. Pigs treat us as equals.

Winston Churchill (attrib.)

***

Average time a cat sleeps each day:
16 hours

Estimated number of kangaroos in Australia:
40–50 million

Maximum speed of a cheetah:
up to 120 kilometres per hour

Number of species of birds in Australia:
about 800

proportion that are unique to Australia:
50 per cent

Amount of water a mature elephant can carry in its trunk:
up to 4–6 litres
amount it drinks each day:
80–160 litres

Average volume that an elephant urinates each time:
6 litres

in a day: **55 litres**

Number of teeth of a turtle: **0**

Weight of a panda at birth: **90–130 grams**

Number of toes on an ostrich: **2**
on an emu: **3**

Number of teeth a dog has: **42**

Lifespan of a mayfly:
about 2 hours

Proportion of genes that humans share with chickens:
60 per cent
with rodents: **88 per cent**

Weight of a male Siberian tiger:

up to 180 kilograms

Number of teeth in a snail:
up to 25,000

CHILDREN AND TEENAGERS

The first half of our lives is ruined by our parents, and the second by our children.

Clarence Darrow, inventor of the board game Monopoly

My son has taken up meditation–at least it's better than sitting around doing nothing.

Max Kauffmann

Worst ages for men to get acne: **18–19 years**
for women: **16–18 years**

Reported number of teenagers killed each year in Australia due to excessive drinking:
about 50

Proportion of today's teenage smokers who will die from smoking-related illnesses:
50 per cent

Height of the average woman aged 12–24 in Australia: **164** centimetres
weight: **58 kilograms**
bust: **88 centimetres**
waist: **71 centimetres**
hips: **97 centimetres**

Proportion of 16–17-year-olds in Australia who say they have smoked cannabis: **39 per cent**
in 1999: **46 per cent**

Proportion of 16–17-year-olds in Australia who say they have drunk alcohol to excess:
20 per cent

Proportion of Australians aged 15–24 who left high school
early and say they are satisfied with their lives:
73 per cent
who finished Year 12 and say the same:
90 per cent

Proportion of Australian boys who go skateboarding or rollerblading:
28 per cent
play electronic or computer games:
82 per cent

Proportion of children aged 12–16 in Britain who have a job:
34 per cent
their average weekly earnings: **$A64**

Proportion of children in Australia aged under 14 who live with two parents:
81 per cent
a lone parent:
19 per cent

Proportion of children aged 10–17 in Australia who receive no pocket money:
46 per cent

Proportion of 20–24-year-olds in Australia who say they have a religious affiliation:

68 percent

Headmasters have powers at their disposal with which Prime Ministers have never been invested.

Winston Churchill,* My Early Life, *1930

***

Proportion of people aged 15 or less in 2009:
in the world: **27.2 per cent**
in the USA: **20.2 per cent**
in Australia: **18.6 per cent**
in Italy: **13.5 per cent**

Ratio of text messages to phone calls for 16–19-year-old mobile phone users:
10 to 1

Proportion of Australian male school pupils aged 5–14 who read for pleasure: **68 per cent**
female: **82 per cent**

Average time spent per week during school time by 5–14-year-olds in Australia watching TV or videos:
11 hours

Number of teenagers aged 15 – 19 who died in motor vehicle crashes in the USA in 2008:
3,500
proportion of these not wearing a seat belt:
75 per cent

Proportion of Australian men aged 18–19 who say they are dependent on alcohol:
9.3 per cent

Proportion of Australian boys aged 12–15 who have used alcohol, tobacco or cannabis:
39 per cent
of Australian girls:
34 per cent

Proportion of boys aged 12–15 in the USA who have used
alcohol, tobacco or cannabis: **25 per cent**
of girls: **25 per cent**

Proportion of Australian boys aged 12–14 who say they smoke daily: **7 per cent**
of girls: **8 per cent**

Proportion of boys in the USA aged 12–14 who say they smoke daily: **3 per cent**
of girls: **4 per cent**

In any given six-month period, the proportion of 9–17-year-olds who are estimated to be suffering from major depression:
5 per cent

Proportion of conceptions in England and Wales by teenagers:

13 per cent

Total number of hours spent in the classroom by children in Australia aged 7–14:
about 8,000

Proportion of British fathers who will not let their children aged 2–15 watch television unsupervised:
45 per cent
of mothers: **38 per cent**

The only reason I always try to meet and know the parents is because it helps me to forgive their children.

Louis Jahannot

***

Estimated number of children worldwide who live on the streets:
between 100 and 150 million

Proportion of Australian teenagers aged 13–17 who see marriage as a sign of success:
57 per cent

Proportion of 12–15-year-olds in Australia who have used marijuana/cannabis: **5.2 per cent**
painkillers/analgesics: **2.0 per cent**
ecstasy: **0.6 per cent**
smoked: **4.3 per cent**
tried alcohol: **35.2 per cent**

Number of anti–depressant prescriptions given to British teenagers aged 16–18 in full-time education in 2003: **140,000**
in 1995: **46,000**

Proportion of teenagers in the USA who would rather date someone who did not smoke:

86 per cent

Proportion of 18–19-year-olds in Australia who have used: marijuana/cannabis: **26.5 per cent**
painkillers/analgesics: **3.2 per cent**
ecstasy: **8.8 per cent**
smoked: **26.3 per cent**
tried alcohol: **92.6 per cent**

Proportion of surveyed backpackers aged 15–19 who say they have sex in Australia without using a condom:
50 per cent
who never use a condom:
25 per cent
who had sex with a new partner in the previous three days:
39 per cent

Proportion of Australian teenage girls who say that oral sex is not 'sex':
33 per cent
who have had oral sex by age 15:
20 per cent
by age 17:
50 per cent

Proportion of young men who have a poor understanding of how bras do up and undo:
40 per cent

Before I got married, I had six theories about bringing up children.

Now I have six children and no theories.

John Wilmot

***

Estimated annual cost of teenage pregnancies in the USA:
about A$10 billion
proportion of US teenagers who become pregnant at least once before they reach the age of 20:
34 per cent
proportion of those that are unintended:
80 per cent
that are to unmarried women:
81 per cent

Number of teenage pregnancies of girls aged under 18 in the UK in 2007:
42,900

Proportion of Americans who have one or more siblings:
85 per cent
five or more: **30 per cent**

Proportion of Australian girls who take part in art or craft activities:

61 per cent

Proportion of Australian children aged 3–12 who have a vision problem:
25 per cent
proportion of these that go undetected:
67 per cent

Proportion of surveyed British schoolchildren who thought that Adolf Hitler was their Prime Minister during World War II:
4 per cent

Proportion of British children aged 12–16 who can correctly spell:
'Hogwarts': **85 per cent**
'David Beckham': **80 per cent**
'William Shakespeare': **30 per cent**
'Jane Austen': **8 per cent**

Proportion of 11-year-old British children who can correctly spell 'thoroughly':
12 per cent

Proportion of Australian females aged 14–17 who drink alcohol and also participate in risky behaviour such as drink-driving:
33 per cent

Proportion of teenagers aged 12–16 in Australia who have a television in their bedroom:
50 per cent

Proportion of childcare in Australia that is provided by grandparents:
33 per cent

Number of teenagers killed each year in Australia due to
excessive drinking:
about 50

Attention span of the average high school student:
20 minutes

Always be nice to your children because they are the ones who will choose your rest home.

Phyllis Diller

***

Proportion of Australian teenagers aged 13–17 who think their future looks bright:
85 per cent
who feel that they are currently not doing well enough in their life:
24 per cent

Proportion of young Australians who say smoking marijuana is relaxing:
60 per cent
who describe it as 'fun':
25 per cent

Estimated number of children who die each day worldwide from huger-related causes:
over 16,000

Proportion of Australian adults who will never have children:
22 per cent

Date that Playschool first aired on television in Australia:

18 July 1966
running time: **25 minutes**

Proportion of Australian teenagers who are in full-time study or full-time work:
85 per cent

SEX

Continental people have sex lives; the English have hot water bottles.

***George Mikes,* How to be an Alien (1912)**

Literature is mostly about having sex and not much about having children.

Life is the other way round.

David Lodge,* The British Museum is Falling Down, *1965

***

Average number of acts of human sexual intercourse in the world each day:
about 120 million

Proportion of British women who would wear no underwear when on a very special date:
25 per cent

Number of kilojoules burned in one minute by vigorous kissing:
0.025
banging your head against a wall for one minute:
0.01

Proportion of surveyed British men who fantasise about having sex with a stranger: **42 per cent**
women: **32 per cent**

Percentage of bird species that are monogamous:
90 per cent

Proportion of surveyed Australian males who say they have visited an internet sex site in the previous 12 months:
20 per cent
of women: **2.5 per cent**

Proportion of surveyed Australians who believe it is easier to access pornography than it was 5–10 years ago:
82 per cent

Number of positions for sexual intercourse:
almost limitless
proportion of couples who have tried more than 20 of them:
5 per cent

Proportion of surveyed sexually active 15–19-year-old teenagers in Australia who do not use condoms consistently:
45 per cent

Proportion of people who feel that the first kiss tells what a longer-term relationship will be like:
80 per cent
who would refuse to date a bad kisser:
90 per cent
who would not kiss on a first date:
33 per cent

Personally I know nothing about sex because I've always been married.

Zsa Zsa Gabor,* Observer, *16 August 1987

***

Proportion of Australian men who have fantasised about having sex with a female co-worker:
80 per cent

Proportion of men who feel it is okay to cheat on a partner if you are not caught:
36 per cent
it is impossible to stay faithful:
11 per cent

Age at which women reach their sexual peak: **35 years**
men: **18 years**

Proportion of adults who have had sex at work:
in Australia: **18 per cent**
in South Africa: **30 per cent (world's highest)**
in Germany: **6 per cent (world's lowest)**

Proportion of Viagra prescriptions in the USA that are to men aged over 40:
98 per cent

Proportion of Australians who have had sex in a park:
54 per cent
South Africans: **30 per cent (highest)**
Thai: **5 per cent (lowest)**

Proportion of Australian men who say they do not enjoy
having sex:
20 per cent

Proportion of British cat or dog owners who send their pet out of the bedroom before they have sex:
89 per cent

Proportion of Canadian women who prefer eating chocolate to having sex:
38 per cent

Proportion of people seeking treatment for sex addiction in Australia who are women:
1 in 3

Proportion of surveyed British people who say they never have sex:
13 per cent

Estimated number of abortions in England and Wales each year:
100,000

Proportion of surveyed German men aged 20–60 who say they have cheated on their partner at least once:
51 percent
proportion who said they did so from a lack of love or interest from their partner:
47 percent

Proportion of Australians aged 16–60 who have never had sex:
8 per cent

I remember the first time I had sex—I still have the receipt.

Groucho Marx

***

Proportion of surveyed German women aged 20–60 who say they have cheated on their partner at least once:
43 percent
proportion who said they did so from a lack of love or interest from their partner:
61 percent

Proportion of adults who say they have sex weekly:
Greece: **87 per cent (world's highest)**
French: **70 per cent**
Indians: **68 per cent**
New Zealand: **63 per cent**
Australians: **60 per cent**
Britons: **55 per cent**
Americans: **53 per cent**
Japanese: **34 per cent (world's lowest)**

Chance of becoming pregnant while using a condom perfectly correctly once a week over a year:
3 per cent
with typical usage:

10 per cent

Proportion of surveyed Australian women aged 12–20 who said a boyfriend had tried to force them to have sex:
14 per cent
who had physically forced them to have sex:
6 per cent

Proportion of new cases of sexually transmitted diseases that reportedly occur in teenagers:
25 per cent

Proportion of Australian men aged 70 and over who are sexually active:
37 per cent
who wished they had more sex:
25 per cent
proportion who have difficulties with erectile dysfunction:
68 per cent

Proportion of virgins in Australia who have had oral sex:
13 per cent

Proportion of surveyed Australian men who say that they have had sex with someone else while in a relationship: **28 per cent**
of women: **22 per cent**

Proportion of teenagers in China who get their information about sex:
from pornography: **70 per cent**
from school: **1.7 per cent**
from parents: **1.3 per cent**

Average time between a man's thoughts about sex:
8 minutes

Proportion of males in Australia who want tougher controls on pornography: **59 per cent**
of females: **80 per cent**

I use to be Snow White but I drifted

Mae West (attrib.)

***

Average volume of sperm ejaculated by males:
aged 20–30 years: **4.0 millilitres**
aged 30–50: **3.5 millilitres**
aged 60–70: **2.0 millilitres**
aged 70 and over: **1.0 millilitres**

Proportion of men who ejaculate within:
one minute of penetration: **21 per cent**
5 minutes of penetration: **62 per cent**

Normal length of a human male's flaccid penis:
8.8 centimetres

Length of a vagina (unstretched):
7–11 centimetres

Proportion of surveyed Australian males who said they have tried to force a girl to have sex:
3 per cent
proportion of these who were successful in having sex with a girl using physical force:
67 per cent

Proportion of Australian adults who have sex on average less that once a week:
60 per cent
among married couples:
69 per cent

Proportion of Australian men who say they would like to have
more sex: **60 per cent**
of women: **70 per cent**

Proportion of males who would be flattered if a female co-worker asked them to have sex:
66 per cent

Proportion of females who would be insulted if a male co-worker asked them to have sex:
66 per cent

Most male sexual partners in one day (US porn star Lisa Sparxxx at Eroticon 2004): **919**
her age at the time: **27 years**
time taken: **about 12 hours**

Most climaxes with female sexual partners in one day (male porn star Jon Dough in the movie *The World's Luckiest Man*): **55**
year: **1997**

Proportion of men who say they would have sex with a female boss just to get a promotion:
45 per cent

Proportion of surveyed Australian women aged 16 and over who enjoy kissing men:
56 per cent
who feel that kissing is better than sex:
60 per cent

Prostitution gives a girl an opportunity to meet people. It provides fresh air and wholesome exercise, and it keeps her out of trouble.

Joseph Heller,* Catch-22, *1961

***

Average number of times couples in Italy have sex each year if there is no TV in their bedroom: **96**
with a TV in their bedroom: **48**
if they have a TV in their bedroom and are also aged over 50: **18**

Proportion of surveyed French men aged over 40 who have a willingness to improve their sexual performance: **60 per cent**
of British men: **46 per cent**

Largest recorded erect penis:
34 centimetres

Proportion of men who are sexually active at age 80:
12 per cent

Proportion of women aged over 40 who stay that staying young is about changing lifestyle, not looks:
49 per cent

Proportion of 70-year-old men who still fantasise about sex:
70 per cent
70-year-old women: **45 per cent**

Proportion of women who have faked an orgasm at least once:
92 per cent

Number of people in Australia in 2008 diagnosed with a sexually transmitted infection (STI):
about 70,000

Proportion of surveyed women in Australia who said they have never experienced orgasm during sexual intercourse:
14 per cent
always experience orgasm:
23 per cent
sometimes experience orgasm:
63 per cent

Proportion of surveyed Australian men aged 16–60 who say they want sex every day: **25 per cent**
women: **8 per cent**

Average age for an Australian male to have sex for the first time: **16**
female: **16**

Proportion of Australian men aged 16–60 who have had more than one sexual partner in the past year: **15 per cent**
women: **8 per cent**

Proportion of men who have extramarital sex that have slight or no emotional involvement: **44 per cent**
of women: **11 per cent**

Time a couple is most likely to have sex: **10.30pm**

Average number of times a day that a rat has sex: **20**
a lion in the mating season: **up to 50**

A promiscuous person is one who is getting more sex than you.

Victor Lownes,* Playboy, *1985

***

Proportion of Australian men aged 16–60 who say that sex with their regular partner is very pleasurable: **90 per cent**
women: **79 per cent**

Average number of sexual partners surveyed Australian males claim to have had: **17**
women: **7**

Proportion of men who say they think about sex several times a day: **54 per cent**
of women: **19 per cent**

Proportion of men who would rather give up sex for a week than their remote control: **9 per cent**
of women: **18 per cent**

Proportion of women who think they are having more sex than their girlfriends:
58 per cent

Proportion of women with a positive self-image in the USA who have orgasms every time they have sex:

73 per cent
of self-conscious women:
42 per cent

Normal range of a human male's erect penis:
12–19 centimetres

Size of a whale's erect penis:
up to 3 metres

Proportion of internet searches that are porn related:
25 per cent

Estimated proportion of all web sites that are pornographic:
12 per cent

Proportion of surveyed couples in an experiment who reached orgasm during intercourse when they kept their socks on:
80 per cent
without socks: **50 per cent**

If it wasn't for pickpockets I'd have no sex life at all.

Rodney Dangerfield

***

Amount of blood required to make a male penis erect:
30 millilitres

Estimated number of women seduced by Casanova (1725–1798):
10,000

According to the Kinsey Report in 1948 the proportion of American men who had been unfaithful to their wives:
56 per cent

Number of women aged 45 and over who say they like having sex:
in the lounge room: **90 per cent**
in the bathroom: **44 per cent**
in the garden: **15 per cent**

Average number of sex partners reported by women:
who were watched as they filled in their survey answers: **2.6**
who were told they were completely anonymous: **3.4**
who thought they were attached to a lie detector: **4.4**

Proportion of Australian men who hope to have sex with more than one partner in the next month: **25 per cent**
of women: **5.8 per cent**

Proportion of women aged 35–44 who say they fantasise about having a one-night stand:
25 per cent

Proportion of married men in Australia who are seeking a 'short-term' relationship: **25 per cent**
of married women: **10 per cent**

Proportion of Australian workers who are currently sleeping with a colleague:
11 per cent

Proportion of men in Australia/New Zealand who say they want more than one sexual partner in the next 30 years: **60 per cent**
of women: **30 per cent**

Increase in the sexual desire of men during a heatwave:
28 per cent
of women: **21 per cent**

Proportion of Australian women who find sex uninteresting or painful:
25 per cent

Length of healthy sperm: **0.05 millimetres**
top speed: **50 microns per second**
time to reach the fallopian tube: **about 1 hour**

Proportion of golfers who say they prefer to play golf than have sex:
11 per cent

I'm a double bagger. Not only does my husband put a bag over my face when we're making love, but he also puts a bag over his head in case mine falls off.

Joan Rivers

***

Proportion of people living in Canberra who say they have:
sex every day: **9 per cent**
no sex at all: **13 per cent**

Proportion of Australian men who say they would like to sleep with somebody famous: **65 per cent**
of women: **35 per cent**

Proportion of women who lie about the sexual performance of their partner:
36 per cent

Proportion of Australian girls aged 14–17 who say they have lost a boyfriend because they refused to have sex:
11 per cent

Proportion of surveyed men who say they have accessed pornography at work: **20 per cent**
women: **13 per cent**

Proportion of French men who are satisfied with the frequency of their sex life: **6 per cent**
French women: **8 per cent**

YEARS OF IMPORTANCE

I've been on a calendar, but never on time.

Marilyn Monroe,* Look, *1962

Put your hand on a hot stove for a minute, and it seems like an hour.

Sit with a pretty girl for an hour, and it seems like a minute. That's relativity.

Albert Einstein

***

Years of marriage for a traditional wood wedding anniversary:
5

Year in which the name 'Australia' (first suggested by Matthew Flinders) was officially recognised:
1824

Year that Vegemite was invented and launched (by Dr C.P. Callister):
1923

Year of the last reported sighting of the Rodrigues Solitaire dodo:
1790

Year that whipping young children was banned in Britain:
1932

Year the drinking straw was invented (by Marvin Stone):
1888

Year that the USA issued its last 3-cent coin:
1889

Year that the last woman (Ruth Ellis, Wales) was hanged in Britain:
1955

Year that Shakespeare wrote his last play (*The Tempest*): **1611**
year he died: **1616**

Year of the first Australian federal election in which voting was compulsory: **1925**
proportion of the electorate who cast a vote in that election:
91.3 per cent

Year that numbering began in the streets of London:
1764

Year the wristwatch was invented (by Louis Cartier, France):
1904

Year that shoelaces were invented:
1790

Year that parking meters were first introduced (in Oklahoma, USA):
1935

Year the corkscrew was first patented (by Reverend Samuel Henshall, England):
1795

Year that SOS was adopted at the Berlin Radio Conference as the international distress signal:
1906

Half our life is spent trying to find something to do with the time we have rushed through life trying to save.

Will Rogers

***

Year that the last convict ship, the *Hougoumont,* from Britain
arrived in Australia:
1868

Year that Jack the Ripper operated:
1888

Year that dynamite was invented (by Alfred Nobel, Sweden):
1867

Year that Ceylon became a republic and changed its name to Sri Lanka:
1972

Year the rubber band was invented (by Stephen Perry, London):
1845

Year that crossword puzzles were invented (by Arthur Wynne, USA):

1913

Year that most women were given the right to vote in Australian federal elections and stand for the national parliament:
1902

Year escalators were invented (by Jesse W. Reno, USA):
1891

Year that Albert Einstein won the Nobel Prize for Physics:
1921

Year that typewriters were invented (by Christopher L. Sholes, USA):
1868

Year that the last western country (Switzerland) granted women the right to vote:
1971

Year the Berlin Wall was erected: **1961**
year it came down: **1989**

Year the printing press was invented (by Johannes Gutenberg):
1455

Year the bra was first patented (in New York): **1913**

Year that the term 'rock and roll' was coined:
1951

Year of the first stolen car (a Peugeot in Paris):
1896

Year that DVDs were launched:
1997

The butterfly counts not months but moments, and has time enough.

Rabindranath Tagore

***

Year that Russian cosmonaut Yuri Gagarin became the first person in space (in *Vostok1*):
1961

Year Patrick White won the Nobel Prize for Literature:
1973

Year that the face of George Washington was completed on
Mount Rushmore: **1930**
Thomas Jefferson: **1936**
Abraham Lincoln: **1937**
Theodore Roosevelt: **1939**

Year the world's first Laundromat opened (in Fort Worth, Texas):
1934

Years that the Italian explorer Marco Polo lived:
1254–1324

Year that the Australian magazine *Cleo,* with a male nude centrefold (Jack Thompson), was launched:

1972

Year that Mickey Mouse first appeared:
1928

Year that John Lennon married Yoko Ono in Gibraltar:
1969

Year that *The Magic Pudding* by Norman Lindsay was first published:
1918

Year the Indian Prime Minister Indira Gandhi was assassinated at age 67: **1984**
Benazir Bhutto, Pakistani politician, was assassinated at age 54: **2007**

Year that a television couple was first portrayed as sharing a bed (in the US sitcom *Mary Kay and Johnny*):
1947

Year that actress Jayne Mansfield posed for *Playboy:* 1956
Elizabeth Taylor: **1963**
Joan Collins: **1969**
Raquel Welch: **1977**
Sharon Stone: **1990**

Year the tunnel under the English Channel opened:

1994

Year of the first tester for alcohol on a person's breath (the Drunkometer, USA):
1939

Year Identikit pictures were first used (in the UK):
1961

In 100 years, all new people.

Stacey Boltz

***

Year of the first interracial kiss on television (between Captain Kirk and Lieutenant Uhuru in *Star Trek*): **1968**
kiss between two women (Lamb and Perkins in L.A. Law): **1991**

Year that Pluto was discovered: **1930**

Year the first PC was marketed: **1972**

Year that Samuel Colt invented the revolver (a 'six shooter'): **1835**

Year that the novel *Gone with the Wind* (by Margaret Mitchell) was first published: **1936**

Year the Dolby anti-hiss sound system was invented: **1968**

Year the full stop was introduced: **c. 765**

the apostrophe: **1510**
the comma: **1534**
the ampersand: **1837**

Year that actor Charlie Chaplin died:
1977

Year the first electric dishwasher was marketed (in the USA):
1932

Year the last person was burnt at the stake (Catherine Murphy at Newgate, London):
1789

Year the cash register was first patented (by James J. Ritty, USA):
1879

Year the Globe Theatre was first founded in London: **1599**

Year that a tax on windows was introduced in England: **1696**
removed: **1851**

I've got all the money I'll ever need, if I die by four o'clock.

Henny Youngman

***

Year Interpol was founded:
1923

Year the Barn Dance originated:
1892

Year the planet Neptune was discovered (by Johann Galle, Germany):
1846

Year chloroform was discovered (by Samuel Guthrie, USA):
1831

Year of the bubonic plague epidemic in London:
1665

Year nylon toothbrushes were first marketed (in the USA):
1938

Year the Red Cross was founded (by Henri Dunant, Switzerland):

1864

Year that *'Silent Night'* (written by Austrians Joseph Mohr and Franz Gruber) was first performed:
1818

Year that Velcro was invented (by George de Mestral, Switzerland):
1948

Year that the foxtrot originated (named after Broadway dance star Henry Fox):
1912

Year in which Australia Day was first celebrated:
1818

Year the first exclamation mark appeared:
1553

Year the Splayd was invented:
1946

Year Australia's first Totalisator Agency Board (TAB) opened (in Victoria):
1961

Year that General Custer made his last stand at the Battle of Little Big Horn: **1876**
his age: **36**

Year the song 'Happy Birthday' (melody by Mildred J. Hill, lyrics by Patty Smith Hill) was written: **1893**
first copyrighted: **1935**

Year the internet began: **1969**
the first email was sent: **1971**

The hardest years in life are those between ten and seventy

Helen Hayes, actress, at age 73

***

Year that the farthing was last used as legal tender in the UK:
1960

Year the *Sydney Morning Herald* commenced publication (as a weekly): **1831**
number of pages: **4**
cost: **7 pence**

Year the Sydney Opera House opened:
1973

Year the stapler was patented (by Samuel Slocum, US):
1841

Year the sewing machine was first patented (by Isaac Singer, US):
1855

Year the tuning fork was invented (by John Shore, England):
1711

its pitch: **A423.5**

Year of the first gas chamber execution (Gee Jong, Carson City, Nevada, for murder): **1924**
of the first lethal injection execution (Charlie Brooks Jr, Huntsville, Texas, for murder): **1982**

ENTERTAINMENT

A celebrity is a person who works hard all his life to become known, then wears dark glasses to avoid being recognised.

Fred Allen

Hollywood is where they shoot too many pictures and not enough actors.

Walter Winchell

***

Length of the 'shower scene' in the original *Psycho* movie:
45 seconds

Number on the door of Jerry's apartment in the TV show *Seinfeld:* **5A**
of Kramer's: **5B**

Number of costumes worn by Elizabeth Taylor in the 1963
movie *Cleopatra:* **65**

Proportion of surveyed Britons who say they enjoy television advertisements:
24 per cent

Average number of television advertisements seen by Australian schoolchildren each year:
25,000

Elvis Presley's army serial number:
53310761

Box office takings in Australia of *New Moon* in its opening weekend in 2009 (a record): **$16.01 million**
number of screens on which it was shown: **530**

Of the ten highest grossing movies in 2009 in the USA, the number that were:
PG rated: **3**
PG–13 rated: **6**
R rated: **1**

Number of people who put their religion as 'Jedi' in the 2001 British National Census:
390,000

Estimated number of Australians who watched the 2005 wedding of Prince Charles and Camilla Parker Bowles on television:
1.46 million

Number of books sold by English author Enid Blyton:
over 600 million
number of languages: **40**
year she died: **1968**

Cost of a 30-second commercial during the final episode of the TV show *Friends:*
$A2.75 million

Year that *Lost in Space* premiered on television: **1965**
year in which it was set: **1997**

Year of the last film to win Oscars for both best actor and best actress (Jack Nicholson and Helen Hunt in *As Good As It Gets*):
1998

Never try to teach a pig to sing; it wastes your time and it annoys the pig.

***Robert A. Heinlein,* Time Enough for Love**

****

Average television viewing time per week in Britain: **18 hours**

Year that the TV show *Coronation Street* began broadcasting in Britain:
1960

Proportion of movie villains who smoke: **36 per cent** heroes: **21 per cent**

Most Oscars won by a film (*Ben-Hur* in 1959, *Titanic* in 1997 and *Lord of the Rings: Return of the King,* 2003): **11**

Approximate number of deaths in the 1990 movie *Die Hard 2:*
264
Rambo 3: **106**
Robocop 2: **74**

Most consecutive No.1 singles on US charts (by Whitney Houston 1995–1988): **7**

Number of movies in which Robert de Niro dies (the highest):
14
Bruce Willis: **11**

Number of spots drawn by animators for the 1996 Disney film *101 Dalmatians:*
6,469,952

Number of official James Bond movies with just one word in the title:
5 *(Octopussy, Goldfinger, Moonraker, Goldeneye, Thunderball)*

Estimated cost to legitimate film business in Australia each year due to video piracy:
$500 million

Number of Harry Potter books sold worldwide by June 2009:
over 400 million

Time all clocks show in the movie *Pulp Fiction:*
4.20

Years in which *Number 96* was Australia's top-rating television show:
1973 and 1974

Worldwide sales of Michael Jackson's album *Thriller* by 2009:
about 110 million

Height of actor Alan Ladd: **165 centimetres**
Danny DeVito: **146 centimetres**

Opera is when a guy gets stabbed in the back and, instead of bleeding, he sings.

Ed Gardner in Duffy's Programme, 1940s

***

Number of rats specially bred for the 1989 movie *Indiana Jones and the Last Crusade:* **3,000**

Size of Olive Oil's (Popeye's girlfriend) shoe: **14A**

Number of times Arnold Schwarzenegger has won the Mr Olympia title: **7**

Year in which Marilyn Monroe was the first *Playboy* magazine centrefold: **1953**

Year that Vivien Leigh won her first Academy Award for Best Actress in the movie *Gone With The Wind:* **1939** her second in *A Streetcar Named Desire:* **1951**

Estimated Australian television viewing audience when Cathy Freeman won her gold medal at the Sydney 2000 Olympics:

6.3 million

Year that *Star Trek* started on television: **1966**

Most Oscar nominations by a film (*All About Eve,* 1950 and *Titanic,* 1997): **14**

Number of camels once offered by an Arab sheikh to Dennis Hamilton in exchange for his wife, actress Diana Dors:
23

Number of *Superman* movies made by Christopher Reeve: **4**
number of television episodes of *Superman* made by George Reeves: **104**

Proportion of British people who regularly watch sport on TV in a pub:
24 per cent

Final year arrived at in H.G. Wells's 1895 novel *The Time Machine:*
802701

Estimated number of Elvis impersonators in the world:
35,000

Number of movie tickets sold annually worldwide:
9 billion

Number of times actor Yul Brynner performed in *The King and I:*
4,625

I don't know anything about music. In my line you don't have to.

Elvis Presley

***

Amount for which Marilyn Monroe's bra sold for at a Sotheby's auction in 1981:
$A1,400
Madonna's panties in 1991:
$A15,000

Proportion of Americans who are vegetarian:
3 per cent

Number of men who made up the Three Stooges: **6**
number who were real brothers: **3 (Moe, Curly and Shemp)**

Number of piglets used in the filming of the 1995 movie *Babe:*
48

Longest Oscar speech (by Greer Garson in 1943):
5 minutes 30 seconds
shortest (Alfred Hitchcock, Best Director in 1958):
2 seconds (he said 'thank you')

Nicole Kidman's star number on Hollywood Boulevard's Walk of Fame:
2211

Number on Sergeant Joe Friday's badge in the TV series *Dragnet:*
714

Number of episodes made of the TV series *The Honeymooners:* **39**

Number of Perry Mason 'Case of the...' novels written by Erle Stanley Gardner: **75**

Number of movies in which actor John Wayne appeared: **153**
number of these in which he had a leading role: **142**

Proportion of Australian adults aged under 35 who:
routinely listen to portable music devices: **45 per cent**
spend time in loud nightclubs or bars: **38 per cent**
frequent stores that blast loud music: **23 per cent**

Number of curtain calls received by Pavarotti at the Berlin Opera in February 1988:
165

Number of radio stations in the US: **about 14,000**
year of the first radio broadcast: **1920**

Proportion of complaints about TV advertising that are in relation to sex or nudity:
30 per cent

Nothing wrong with shooting as long as the right people get shot.

Clint Eastwood

***

Average number of books Americans say they read each year:
16

proportion who say they read no books in the previous year:
17 per cent

Proportion of worldwide individual visits to the movies that take place in India:
45 per cent

Amount for which James Bond's 1965 Aston Martin DB5 sold at auction in January 2006 in the USA:
$A2.8 million dollars

Estimated proportion of all music disks sold worldwide that are pirated:
1 in 3

Proportion of Australian women who want full-figured women in advertisements:
72 per cent

less airbrushing in photographs:
79 per cent

Number of Australians who watched the penalty shootout in the Australia v Uruguay World Cup soccer qualifier in 2006:
3.4 million

Auction price paid in July 2003 for a costume worn by television's *Superman,* George Reeves:
$200,000

Dirty Harry's badge number:
2211

Number of years for which Jack Mercer was the voice of Popeye the Sailor:
45 (1935–1980)

Number of days the painting *Le Bateau* by Henri Matisse was inadvertently hung upside-down at New York's Museum of Modern Art in 1961: **46**
number of visitors who failed to notice: **116,000**

Number of Disney characters that have reportedly been given official birthdays:
3 (Mickey, Minnie and Donald)

Number of his own paintings burned by artist Monet in 1866 to prevent them being seized by creditors:

about 200

Men don't care what's on TV. They only care what else is on TV.

Jerry Seinfeld

***

Amount lost in 2009 by players in legal gambling in the USA:
about US$100 billion

Number of animals that appeared in the 1956 movie *Around the World in 80 Days:* **8,552**
number of sheep: **3,800**
buffalo: **2,448**
donkeys: **950**
horses: **800**
monkeys: **512**

Age of actress Joan Collins when she posed semi-nude for *Playboy* magazine in 1983:
50 years

Number of extras used in the 1981 film, *Ghandi* (for a 10 minute funeral scene):
300,000

Length of film in a 2-hour movie:
3,292 metres

Number of movies portraying Zorro: **69**

Age of Orson Welles when he co-wrote, produced, directed and starred in the 1941 movie, *Citizen Kane:* **25 years**

Length of the sword fight sequence between Stewart Granger and Mel Ferrer in the 1952 film, *Scaramouche* (the longest): **6 minutes 30 seconds**

Age at death of Goons' member Sir Harry Secombe: **79 years**
year: **2001**

Chance of rolling a total of seven with two dice: **1 in 6**
a total of eleven: **1 in 18**
a total of seven or eleven: **1 in 4.5**

Diameter of a compact disc: **12 centimetres**

Most Oscars won (Walt Disney, from 64 nominations): **26**

Height of an Oscar: **34 centimetres**
weight: **4 kilograms**

Estimated number of people who queued in Melbourne in 1954 to see an appearance by US cowboy 'Hopalong Cassidy' (actor William Boyd):
15,000

Forget it Louis. No Civil War picture ever made a nickel.

Irving Thalberg,* MGM production executive, advising Louis B. Mayer not to bid for the film rights of *Gone With The Wind, 1936

***

Year that actress Grace Kelly married Prince Rainier II of Monaco:
1956
number of guests at the ceremony: **1,200**

Proportion of the US television viewing audience of 54 million who watched Elvis Presley first sing on the *Ed Sullivan Show:*
82 per cent
year: **1956**

Number of viewers of the most watched television show in Australia in the 21st century (final of the men's Australian Open tennis final, 2005):
4.045 million
second most watched (Rugby World Cup final, 2003):
4.016 million

Year *The Simpsons* first aired on television as a series in the USA: **1989** number of episodes made by 2010: **about 450**

Number on Frasier Crane's apartment door:
1901

Number of keys on a piano: **88**

Number of women who appeared in the Stanley Kubrick movie *Dr Strangelove:* **1 (actress Tracy Reed as Miss Scott)**
year: **1964**

Number of Von Trapp children in the movie *Sound of Music:* **7**
year: **1965**

Length of time *Robinson Crusoe* was marooned on his desert island:
28 years

How long Yorrick's skull was buried before being found by Hamlet:
23 years

Year that Dorothea Mackellar's poem 'My Country' was first published in the *Spectator:*
1908

Year that Neville Shute's novel *A Town Like Alice* was first published in Melbourne and London:

1950

Number of honorary doctorates awarded to actor Bob Hope: **58**

You'd better learn secretarial work, or else get married.

Emmeline Snively, **director of the Blue Book Modelling Agency, to Marilyn Monroe, *1944***

***

Number of admissions to Australia's museums and art galleries in the year ending 30 June 2008:
over 30.7 million
proportion that were free:
68 per cent
number of museum objects/artworks held:
52.5 million
proportion of objects/artworks on public display:
5.4 per cent

Number of numbered symphonies by the composer Joseph Haydn:
104

Largest one-day gross takings by a movie in the US (*The Twilight Saga: New Moon* on 20 November 2009):
US$72.7 million

Time it took to film the movie Paranormal Activity:
10 days
year: **2007**

Longest time spent watching TV continuously (Suresh Joachim in Stockholm): **72 hours**
year: **2009**

EVENTS AND PLACES

She said that all the sights in Rome were named after London cinemas.

Nancy Mitford,* Pigeon Pie, *1940

Nobody goes there anymore because it's always so crowded.

Yogi Berra

***

Size of the numerals on Big Ben: **61 centimetres**
the minute hand: **4.27 metres**

Height above mean sea level of the top of the arch of the Sydney Harbour Bridge:
134 metres
length of the arch span:
503 metres

Land area of Singapore (including about 60 islets):
699 square kilometres

population density per square kilometre in 2010:
6,680

Number of people on board the *Endeavour* when it set sail from England in August 1768: **94**
duration of the round trip voyage: **1,051 days**

Width of sleeping space allowed convicts on board ships from England to Australia after 1802:
45 centimetres

Land area of Europe:
about 10.4 million square kilometres

Average lifespan of a blood cell:
120 days

Height to which the first manned helicopter rose (in 1907):
60 centimetres

Number of islands in the Philippines archipelago:
7,107

Number of provinces in Pakistan:
4 (Punjab, Sindh, North-West Frontier, Baluchistan)

Proportion of surveyed Americans aged 18–24 who could not find England on a world map: **60 per cent** the USA: **11 per cent**

Go to Heaven for the climate, Hell for the company.

Mark Twain

***

Population of Malaysia in 2010:
about 25.7 million

Number of surviving original copies of the final *Inspectimus* issue of the Magna Carta, produced in 1297: **4**
year the charter was initially proclaimed by King John: **1215**

Population density per square kilometre of land in Bangladesh in 2009 (most crowded country): **1,152**
Australia: **2.6**

Length of the Great Wall of China: **6,700 kilometres**
average height: **10 metres**
average width: **5 metres**

Number of people who signed the US declaration of independence on 4 July 1776: **2**

Number of islands in the Bahamas: **about 700**
number that are inhabited: **30**

Number killed in world's worst single aircraft accident (near Ueno, Japan in 1985): **535**
number who were passengers: **520** crew: **15**
number of survivors (all female): **4**

Proportion of the world's seawater that is in the Pacific Ocean:
48 per cent
Atlantic Ocean: **23 per cent**

Total land area of New Zealand:
about 268,000 square kilometres

Number of the world's countries that are located directly on the Equator: **12**

Number of wives of King Henry VIII: **6**

Proportion of the Earth's surface covered by tropical rainforests:
7 per cent

Proportion of the world's forests that are in Siberia:
more than 25 per cent

Number of member states of the United Nations:
192

Number of active volcanoes in Indonesia: **about 400**
average number of earthquakes each day: **3**

Number of US states with a one-syllable name:
1 (Maine)

Length of the Berlin Wall: **about 155 kilometres**
number of watchtowers: **302**

No, I'm from Iowa. I only work in outer space.

***Captain Kirk,* Star Trek IV: Voyage Home**

***

Number of steps in the Eiffel Tower:
1,665

Surface area of the Pacific Ocean:
about 166 million square kilometres
average depth:
3,940 metres

Area of the Island of Nauru in the South Pacific Ocean:
21 square kilometres
population in 2010:
about 13,770
median age:
20 years

Number of countries in which the national flag is not the same
on both sides:
1 (Paraguay)

Area of Vatican City: **0.44 square kilometres**
population: **about 925**

Number of people who reached the summit of Mt Everest on the date 10 May 1993: **40**

Number of men to have become both Vice President and President of the USA without being elected to either position:
1 (Gerald Ford)

Number of visitors to the World Exhibition in Paris in 1900:
50 million

Number of books written about Abraham Lincoln: **about 1,100**
George Washington: **about 750**

Height of the world's tallest dam (Nurek in Tajikistan):
300 metres
year it was completed: **1980**

Distance of the Alice Springs–Darwin railway: **1,420 kilometres**
year of completion: **2004**

Percentage of Portugal that is forest: **40 per cent**
Swaziland: **30 per cent**

Number of routes (on-ramps and off-ramps) on the Orange Crush Interchange in California (the world's most complex): **34**

number of lanes: **66**
number of bridges: **13**

Weight of the Statue of Liberty: **204.5 tonnes**
length of its extended right arm: **12.8 metres**
of its nose: **1.48 metres**
weight of a fingernail: **1.5 kilograms**

When I was a kid my parents moved a lot, but I always found them.

Rodney Dangerfield

***

Total area of Australia:
7,686,850 square kilometres

Proportion of the world's population that speaks, either as a primary or secondary language:
English: **16 per cent**
Mandarin: **16 per cent**
Hindi/Urdu: **15 per cent**
Spanish: **7 per cent**

Strength on the Richter scale of the strongest recorded earthquake (Chile, 1960): **9.5**

Distance from Tasmania to Victoria at their closest point:
240 kilometres

Population of the Galapagos Islands: **about 40,000**
number of large islands: **13**
small islands: **5**

Amount paid at a charity auction in the Gulf State of Qatar in 2006 for the mobile telephone number 666–6666:
US$2.75 million

Proportion of the world's fresh surface water that is contained in Lake Baikal in Siberia:
20 per cent
its depth below sea level:
1,620 metres

Number of hours it takes to mow the lawn at the White House, USA: **8**
height of the grass: **7.5 centimetres**

Length of the river Nile:
about 6,695 kilometres

Height of the Arc de Triomphe de l'Etoile in Paris:
49.5 metres

Number of floors in the Empire State Building in New York: **102**
building height to the roof: **381 metres**
to the top of the antenna: **448.7 metres**

Number of people who fainted in the heat at the opening of the Sydney Harbour Bridge on 19 March 1932:
about 2,500

Distance travelled by Charles Lindbergh, the first non-stop solo flight between New York and Paris in 1927: **5,800 kilometres**
time: **33 hours 39 minutes**
average speed: **173 kilometres per hour**

There is nothing good to be had in the country, or if there is, they will not let you have it.

William Hazlitt,* The Round Table, *1817

***

Number of firemen who fought the blaze that destroyed the 85-year-old Crystal Palace in South London in 1936:
500

Diameter of the satellite Sputnik–1 launched by Russia on 4 October 1956: **56 centimetres**
weight: **83 kilograms**
speed: **29,000 kilometres per hour**
time it took to orbit the Earth: **98 minutes**

Highest temperature ever recorded on Earth (achieved in plasma created in the 'Z machine', in New Mexico, USA):
2 billion degrees centigrade
year: **2006**

Longest earthquake ever recorded (Sumatran-Andaman on 26 December 2004):
500–600 seconds
magnitude: **8.7**
equivalent bomb energy released:
100 gigatonnes

length of the gash in the Earth's seabed:
1,300 kilometres

Length of the *Hindenburg* (the German built zeppelin):
245 metres
maximum diameter: **41 metres**
weight: **220 tonnes**
number of passengers killed when it caught fire in 1937: **35**
number who survived: **62**

World's longest stretch of straight railway track (across the Nullarbor Plain, Australia):
478 kilometres

Length of the 1896 war between Britain and Zanzibar:
38 minutes

Of all the gin joints in all the towns in the world, she walks into mine.

Humphrey Bogart,* Casablanca, *1942

***

Depth of the Dead Sea below sea level (the lowest point on the Earth): **417 metres**
average full days of sunshine it receives each year: **330**

Retail floorspace of North America's largest shopping mall
(West Edmonton Mall, Alberta, Canada): **350,000 square metres**
number of stores: **over 800**
parking spaces: **20,000**
employees: **over 23,000**

Length of the Todd River in the Northern Territory: **315 kilometres**

Distance travelled in its first year of 1928 by the Australian Flying Doctor Service: **32,000 kilometres**
number of flights: **50**
power of the transmitter: **50 watts**

Length of the 'dingo fence' in Central Queensland, Australia:

about 5,600 kilometres
height: **1.8 metres**

Proportion of the world's population that live in urban areas:
48 per cent

Number of books in the US Library of Congress in 2010:
about 21.2 million
the British Library:
about 14 million

Proportion of Australians who live within 85 kilometres of the coastline:
85 per cent

Proportion of Americans who live within 80 kilometres of where they grew up:
50 per cent

Area of the Atlantic Ocean:
76.762 square kilometres

Population of Russia in June 2010:
about 142 million

Length of Iceland's coastline: **4,970 kilometres**
number of countries that border it: **0**

Length of Uluru (the world's biggest monolith, Northern Territory, Australia): **3.6 kilometres**
maximum width: **2.4 kilometres**
circumference: **9.4 kilometres**
height: **348 metres**

Number of beaches in Australia:
about 7,000

PEOPLE

I love mankind – it's people I can't stand.

***Charles Schultz,* Peanuts**

I have noticed that people who are late are often so much jollier than the people who have to wait for them.

Robertson Davies

***

Approximate years of birth of the 'silent' generation:
before 1946
baby boomers: **1946–1964**
Generation X: **1965–1977**
Generation Y: **1978–1989**
Generation Z: **1995 onwards**

Proportion of people who say they dislike shopping for groceries:
67 per cent

Number of US patents of Thomas Edison:
1,093

Number of bachelors who have been elected US President:
2 (James Buchanan in 1857, Grover Cleveland in 1885)

Reported emails received each day by Bill Gates:
about 4 million

Number of American presidents to have been re-elected after being voted out of office:
1 (Grover Cleveland)

Number of people who visit Australian museums each year:
3.6 million
age group with the largest attendance rate:
35–44 years

Number of fans who turned up at Sydney airport in 1956 to welcome US singer Johnny Ray:
4,000

According to the World Health Organisation, the number of overweight people in the world: **about 1 billion**
by 2015: **1.5 billion**

Proportion of pet owners in the USA who share their bed with their pets:
18 per cent

People are needed, but nobody is necessary.

Paul Brown

***

Proportion of people who believe in love at first sight: **53 per cent**

Number of French Kings named Louis: **18**

Number of self-portraits by Rembrandt: **about 100**

Height of former French president Charles de Gaulle: **193 centimetres**

Height of former French emperor Napoleon I: **157 centimetres**

Proportion of people who have complained about a firm or organisation in the last twelve months:
in Sweden: **41 per cent (the world's highest)**
in Britain: **36 per cent**
in Taiwan: **1 per cent**

Proportion of the Swedish population who are blonde: **50 per cent**

Proportion of surveyed builders in the UK with short-cropped hair:
45 per cent

Proportion of women in England and Wales who reach 40 without having children:
20 per cent

Proportion of Australians who sleep at least eight hours a night:
73 per cent

Years of the reign of King Ethelred II (the 'Unready') of England:
978–1013

Height of
George W. Bush: **181 centimetres**
Bill Clinton: **188 centimetres**
George Bush Snr.: **188 centimetres**
Ronald Reagan: **185 centimetres**
Barack Obama: **187 centimetres**

Age of the youngest elected US President:
43 (John F. Kennedy)

Number of asses kept by Roman Emperor Nero's second wife, Poppea, to provide milk for her bath:
500

Length of the reign of Queen Victoria: **64 years**
age at death: **82 years**

Television is more interesting than people. If it were not, we should have people standing in the corners of our rooms.

Alan Coren

***

Number of his own teeth George Washington had when he
died: **2**
Stalin: **3**

Year in which the world's population first reached:
1 billion: **1804**
2 billion: **1927**
3 billion: **1960**
4 billion: **1974**
5 billion: **1987**
6 billion: **1999**
6.5 billion: **2006**
7.0 billion: **2012** (estimated)

Proportion of the world's population with a post-secondary education:
1 per cent
who are unable to read:
70 per cent

Number of axe blows required to completely sever the head of Mary, Queen of Scots in 1587: **3**

Number of Nobel prize winners who were born on each of 21 May or 6 June: **7**

Height of Charlotte Bronte, the author of *Jane Eyre:* **145 centimetres**

Number of surnames to have been held by both a British Prime Minister and a US President: **1 (Wilson)**

Age of Charlie Chaplin's bride, Lita Grey, at their wedding in Mexico in 1924: **16 years**
Chaplain's age: **35 years**

Year that Harry Houdini died (of a burst appendix): **1926**
his age: **52 years**

Year the Australian pathologist Professor Howard Florey was awarded the Nobel Prize for Medicine: **1945**

Year that playwright George Bernard Shaw died: **1950**
his age: **94 years**

Number of Australians in the UK at any one time: **about 300,000**

Wherever I go in this country, people know there is a problem.

Billie Snedden, campaigning for the 1974 Australian Federal election

***

Reported total number of fingers of Anne Boleyn (wife of Henry VIII): **11**

Age of Henry III of England when he succeeded to the throne:
9 years

Age of William Pitt (the Younger) when he first became Prime Minister of Great Britain:
24 years

Estimated number of wigs kept by Queen Elizabeth I:
about 80

Age of Oliver Hardy when he died of a stroke in 1957:
65 years
number of times his partner Stan Laurel appeared in films after Hardy's death:
0

Number of crew of Lord Nelson's flagship, *HMS Victory,* at full strength:

850

Length of the reign of Pope Stephen II, elected in 1952:
3 days

Number of words on the first gramophone record (by Thomas Edison): **5 *(Mary had a little lamb)***
year: **1877**

Year that 'God Save the Queen' (composed by Henry Carey) was first sung:
1740

Last measurement of the world's longest female beard (Vivian Wheeler, USA): **27.9 centimetres**
year: **2000**

Proportion of genes that humans have in common with chickens: **60 per cent**
mice: **88 per cent**

Number of times Julius Caesar was reportedly stabbed:
23

Proportion of people who believe that jealousy can ruin even the best relationships:
83 per cent
who find mild jealousy flattering:
49 per cent

When women go wrong, men go right after them.

Mae West, She Done Him Wrong, 1933 film

***

Australia's population in 2010: **21.5 million**
in 1921: **5.44 million**

Heaviest male in recorded medical history (Jon Brower Minnoch, 1941–1983): **635 kilograms**
height: **185 centimetres**
year: **1978**

Number of US Presidents who had their wedding in the White House:
1 (Grover Cleveland)
year: **1886**
who were divorced at the time of their election:
1 (Ronald Reagan)

Median age of the population of Australia in 2009: **37.3 years** of the world: **28.4 years**

Number of telephones in the world in 1926: **27,750**
proportion that were in the USA: **61 per cent**

Height of the shortest president of the USA (James Madison):
163 centimetres

Proportion of Australian residents born in the UK:
5.5 per cent

Number of English kings named Edward: **8**
Henry: **8**

Number of people reported missing in Australia each year:
about 30,000
proportion who are found (mostly within a month):
99.5 per cent

Number of British sovereigns who have been crowned at Westminster Abbey: **38**

Number of people who cross international borders each day:
about 3 million

Number of passengers (excluding crew) on board the Pilgrim ship *Mayflower:* **102**
number who were called 'John': **10**
year: **1620**

Number of Australians who are homeless on any given night:
about 100,000

Number of left-handed presidents to have served two full terms:

1 (Bill Clinton)

People shouldn't be treated like objects. They aren't that valuable.

P.J. O'Rourke,* Modern Manners, *1984

***

Median age of the US population in 2009: **36.7 years**
in 1900: **22.9 years**

Proportion of Australia's adult population who read for pleasure most days of the week:
78 per cent
who regard books as offering good value for money:
57 per cent

Proportion of Australian adults who smoke: **19.5 per cent**
average number of cigarettes they smoke each day: **5.2**

Proportion of Australia's population aged over 16 who own or use a mobile phone in 2009:
85 per cent

Age at death of singer Ritchie Valens (in 1959): **17 years**
of singer Patsy Cline (in 1963): **30 years**
of singer Janis Joplin (in 1970): **27**

MEDICINE

My doctor is wonderful. Once in 1955 when I couldn't afford an operation, he touched up the X-rays.

Joey Bishop

Doctors think a lot of patients are cured who have simply quit in disgust.

Don Herold

***

Proportion of people with symptoms of bowel cancer who ignore them:
33 per cent

Number of muscles moved with each step taken:
about 54

Number of times healthy young men break wind each day:
14–25
young women: **7–12**

Proportion of adolescent girls affected by scoliosis:
2.5 per cent
proportion of scoliosis-affected people who are female:
90 per cent

Proportion of Australians who think cancer is contagious:
3 per cent
in 1964: **20 per cent**

Proportion of Australian adult males with high blood cholesterol: **8.8 per cent**
females: **7.9 per cent**

Proportion of Australians with a body mass index (BMI) of more than 40 (severely obese): **1 per cent**
of Americans: **5 per cent**

Number of people in Australia diagnosed with coeliac disease:
about 200,000
proportion who are female: **90 per cent**

Proportion of a person's body weight that is blood:
7 per cent

Number of Australians who have arthritis or a musculoskeletal condition:
6 million
proportion for whom it limits their daily activities:
20 per cent

Ten year survival rate of women who are diagnosed with breast cancer:
before the age of 40: **49 per cent**
aged 40–69: **73 per cent**
aged over 69: **63 per cent**

Weight of aspirin used by Americans each year:
16.25 tonnes

Proportion of skin ageing that is due to the sun:
80 per cent

After two days in hospital I took a turn for the nurse.

W.C. Fields

***

Proportion of Australian adult males with high blood pressure:
12.8 per cent
females: **14.7 per cent**

Estimated amount of medicines that find their way into Australia's waterways and landfill each year:
200 tonnes

Proportion of hospital admissions in Australia caused by taking pills:
3 per cent

Number of organ donors in Australia in 2009: **247**
number of people who were on the transplant waiting list on 1 January 2009: **1,716**

Number of Australians each year who suffer from pneumococcal pneumonia:
23,000–36,000

Expected increase in melanoma in Australian men between 2002 and 2011: **28 per cent** in women: **23 per cent**

Proportion of Australians who will need donated blood at some time in their life: **1 in 3**
proportion who currently donate blood: **1 in 30**

Proportion of all deaths in Australia for which the underlying cause is:
cancer: **28 per cent**
the circulatory system: **37 per cent**

Proportion of organ and tissue donors in Australia who are male:
40 per cent

Number of assisted reproductive therapy (such as IVF) babies born in Australia since 1980:
over 80,000
as a proportion of current births:
3 per cent

Most common age to experience a mid-life crisis:
38–44 years

Survival rate of people in London who suffer cardiac arrest:
2 per cent
in other parts of the UK:
10 per cent

Proportion of doctors in China who smoke:
57 per cent

You go to a psychiatrist when you're slightly cracked and keep going until you're completely broke.

Anon.

***

Proportion of males who suffer from sleep apnoea: **4 per cent** of females: **2 per cent**

Number of different bones in an adult skull (excluding the ears): **22**

Number of light-sensitive cells in the retina:
over 100 million

Proportion of 60-year-old men who regularly snore:
60 per cent
of 60-year-old women:
40 per cent

Number of sweat glands each person has:
3–4 million
amount of sweat they are capable of producing in 24 hours:
12 litres

Time it takes for a toenail to grow from toe to tip:
12–18 months

Number of Australians each year who contract Chlamydia:
about 55,000

Number of tastebuds on a human tongue:
about 10,000

Volume of methane in a person's gut at any one time:
about 200 millilitres

Number of times a human blinks each year:
5–12 million

Average circumference of an adult fully erect penis (at the mid-shaft):
12–13 centimetres

Average number of cells that die in your body every minute:
about 300 million

Width of an average human brain: **140 millimetres**
length: **167 millimetres**
height: **93 millimetres**

Number of neurons in a human cerebral cortex:
25 billion

Chance of an Australian man developing cancer by age 75:

1 in 3
an Australian woman: **1 in 4**

God heals and the Doctor takes the fee.

Benjamin Franklin

***

Number of nerve fibres in each human optic nerve:
about 1.2 million

Proportion of cosmetic surgery operations in Britain that are performed on people aged over 60:
8 per cent
proportion of these that are for eyelid reduction (most popular):
38 per cent
proportion of these that are for a facelift:
28 per cent

Average weight of a human head:
3.6 kilograms

Average area of human skin: **2 square metres**
weight: 2.7 kilograms

Speed at which information travels along human nerves:
up to 440 kilometres per hour

Length of nerves in each square centimetre of human skin:

3.42 metres

Number of decibels in a loud snore: **69**

Weight of a human spinal cord: **35 grams**
average length for men: **45 centimetres**
for women: **43 centimetres**

Proportion of Australian males who will develop melanoma:
1 in 25
of females: **1 in 38**
number of people who died in 2007 as a result: **1,279**

Proportion of surveyed Australians who did not know that heart disease was the leading cause of death in women:
97 per cent

Number of new cases of breast cancer in women recorded in Australia in 2010: **about 15,440**
proportion who are aged under 50: **30 per cent**

Amount of urine produced by the average person in a lifetime:
about 50,000 litres

Number of PBS prescriptions filled in Australia in 2007–8:
170.5 million

average cost per filled prescription:
$36.58
annual cost of the PBS:
$6.2 billion

Never go to a doctor whose office plants have died.

Erma Bombeck

***

Number of chromosomes of a crayfish: **200**
a pea: **14**

Proportion of total vision lost if you go blind in one eye:
about 20 per cent

Estimated number of new and recurrent strokes in Australia in 2010:
about 60,000
proportion of these that are to people for the first time:
70 per cent
proportion of these who will die within a month:
20 per cent
proportion of these who will die within a year:
33 per cent

Time it took Napoleon's surgeon, Baron Dominique Larrey (1776–1842), to amputate a man's leg: **2 minutes**
an arm: **30 seconds**

Average weight gained by young Australian women each year:
650 grams
middle-aged women: **500 grams**

Heaviest recorded normal human brain (a 30-year-old male):
2.3 kilograms
lightest (a 46-year-old male):
680 grams
average brain:
about 1.4 kilograms

Weight of a human liver: **1.2–1.5 kilograms**
length: **up to 22 centimetres**
width: **10 centimetres**

Number of cones in each eye (to help see colour):
about 6 million

Number of pain receptors in a human brain: **0**

Year the stethoscope was invented (by René Theophile Laennec, France):
1816

Length of the small intestine of an adult: **6–8 metres**
large intestine: **about 1.5 metres**

Proportion of melanomas that can be cured if treated early:
95 per cent

Proportion of all living organisms in the world that are bacteria:
about 70 per cent

First the doctor told me the good news: I was going to have a disease named after me.

Steve Martin

***

Proportion of Britons who believe what their doctor tells them:
91 per cent

Usual number of bones in a human skull: **28**
hand: **27**

Number of pairs of spinal nerves joining the spinal cord to the rest of the body: **31**

Number of teeth in a human mouth: **32**

Proportion of Australians who are affected by headaches:
55 per cent

Estimated number of people worldwide in 2009 with HIV/AIDS:
33.4 million
number who died:
about 2 million
number of new cases in 2008:
2.7 million

Number of men in Australia each year diagnosed with gonorrhoea: **about 5,000**
of women: **about 2,200**

Chance of a prostate cancer diagnosis in Australia for a male in his: forties: **1 in 1,000**
fifties: **12 in 1,000**
sixties: **45 in 1,000**
seventies: **79 in 1,000**

Normal range of testosterone levels (in nanograms/decilitres) for males in their: forties: **251-914**
fifties: **216–876**
sixties: **196–859**
seventies: **156–818**
age at which levels peak: **17 years**

Proportion of Australia's medical specialists who regularly access online medical databases: **86 per cent**
of GPs: **62 per cent**

Proportion of men who grind their teeth while sleeping:
10 per cent

Amount of liquid the average human bladder can hold:
500–800 millilitres
volume at which there is a desire to urinate:

150–300 millilitres

I'm not feeling very well, I need a doctor immediately. Ring the nearest golf course.

Groucho Marx

***

Proportion of British women who have had cosmetic surgery or say they plan to:
25 per cent
proportion of these who also plan to take out a loan to pay for it:
15 per cent

Number of PBS scripts in Australia written for antidepressants:
in 2007: **20 million**
in 1998: **8 million**
in 1990: **5 million**

Number of muscles used when a human swallows: **25**

Amount of blood a human heart pumps in an average lifetime:
about 200 million litres

Number of cases of testicular cancer diagnosed in Australia in 2009:
about 675
cure rate if discovered early:

95 per cent
age group most likely to develop it:
18–39 years

Proportion of Australians who:
have at least one factor of heart disease: **90 per cent**
are overweight: **60 per cent**
not active enough: **54 per cent**
have high cholesterol: **51 per cent**
have high blood pressure: **30 per cent**
smoke daily: **20 per cent**
drink to dangerous levels: **10 per cent**

Number of new prostate cancer cases diagnosed in Australia each year: **about 20,000**
number who die annually from prostate cancer:
about 3,300

Proportion of people who can flare their nostrils:

30 per cent

My doctor gave me six months to live, but when I couldn't pay the bill he gave me six months more.

Walter Matthau

***

Number of Australians diagnosed with diabetes each year:
about 100,000
proportion that are Type 2 diabetes:
85–90 per cent

Number of pairs of chromosomes in a human: **23**
in a dog: **39**

Average weight of a human kidney: **140 grams**
average length: **10 centimetres**
width: **6.5 centimetres**
thickness: **4.0 centimetres**

Average number of times a woman ovulates in her lifetime:
about 400

World's largest tumour removed intact (a multicystic mass from an ovary in the USA, 1991):
137.6 kilograms

Number of muscles in a human ear: **9**

Average number of times a person flexes their finger joints in their lifetime:
25 million

Proportion of women who are obese in:
the USA: **34 per cent**
Britain: **22 per cent**
Australia: **20 per cent**
France: **11 per cent**
Japan: **3 per cent**

Length of time a vasectomy takes:
15–30 minutes

Proportion of children in Britain with a recognised mental disorder (from depression to autism):
10 per cent

Proportion of smokers in Britain who do not tell their doctor they smoke:
16 per cent

Proportion of people who have 20–20 vision:
33 per cent

Proportion of oxygen entering the bloodstream that is used by the brain:
30–40 per cent

She got her looks from her father: he's a plastic surgeon.

Groucho Marx

***

Proportion of people worldwide who die from:
ischemic heart disease: **12.7 per cent**
cerebrovascular disease: **9.6 per cent**
lower respiratory infections: **6.8 per cent**
HIV disease: **5.1 per cent**

Proportion of 70-year-olds that need glasses: **90 per cent**

Proportion of Lupus sufferers who are female: **90 per cent**

Proportion of people needing a heart, liver or lung transplant in Australia who reportedly die waiting: **20 per cent**

Number of vertebrae in a human spine (including the coccyx as four bones): **33**

Thickness of a human cerebral cortex: **2.0–4.0 millimetres**

Number of babies born each year worldwide who die within the first month of life: **about 4 million**
in the first five years: **about 10 million**

Number of legs on a head louse: **6**

Age at which a human has the most brain cells: **2 years**

Proportion of Australian males that are colourblind: **8 per cent**
females: **0.4 per cent**

Number of Australians with a hearing aid: **about 400,000**

Lifespan of a human taste bud: **7–10 days**
proportion lost by age 60: **50 per cent**

Average weight of skin a human sheds by age 70:
18 kilograms

Area of an adult human's skin: **1.3–2.0 square metres**
minimum thickness: **0.5 millimetres (eyelid)**
maximum thickness: **4 millimetres (palm of hand)**

Number of fat cells in an average adult: **40–50 billion**
an overweight adult: **75 billion**
a severely obese adult: **300 billion**

Average life of an eyelash:
about 5 months

Doctors are the same as lawyers; the only difference is that lawyers merely rob you whereas doctors rob you and kill you too.

Anton Chekhov

***

Number of breaths an adult takes each day:
20,000–23,000

Time it takes blood to do one complete circuit of the human body:
about 1 minute

Amount of air humans breathe in and out each time:
about 500 millilitres
average number of times per minute: **15**

Proportion of surgeons in Australia who are aged over 55 years: **50 per cent**
under 40 years: **16 per cent**

The amount spent annually by Japanese on non-prescription and herbal remedies:
about $A14 million

Longest recorded tapeworm found in a human body:
33 metres

Full capacity of an adult human lung: **4–6 litres**

Proportion of Australians who say they suffer from:
back pain: **49 per cent**
stress: **31 per cent**
lack of exercise: **2 per cent**

Average length of a human male spinal cord: **45 centimetres**
female: **43 centimetres**

Speed at which information travels along a large nerve fibre in a human body:
up to 100 metres per second

Proportion by which the brain decreases by volume and weight after the age of 40 years:
5 per cent

Number of deaths worldwide from measles:
in 2008: **197,000**
in 1999: **871,000**

Average amount of dust a person breathes in during their lifetime:
about 18 kilograms

It is a scientific fact that if you shave your moustache, you weaken your eyes.

William 'Alfalfa Bill' Murray*, Governor of California, *1932

***

Number of different classifications of head or facial pain:
150

Number of people killed annually by the tsetse fly:
about 150,000

Largest gallstone removed (from an 80-year-old woman in London):
6.29 kilograms

Estimated deaths worldwide from the 1918–1919 influenza epidemic:
40–50 million

Year that Medibank was introduced in Australia:
1975

Average number of germs harboured per square centimetre:
on a telephone: **4,020**
on a computer keyboard: **527**

on a computer mouse: **268**

Number of types of haemophilia:
2 (A and B)
proportion of these that is A:
75–80 per cent

Number of the cranial nerve that is paralysed in sufferers of Bell's Palsy: **7**

Number of different chemical components in an average cup of coffee:
over 1,000

Year that Bayer marketed the first drug in tablet form (water soluble aspirin in the USA):
1900

Number of red blood cells in your body at any one moment:
about 2.5 trillion
their lifespan:
120 days

Weight of bacteria in a human gut:
about 1 kilogram

Number of light sensitive cells in the eye's retina:
137 million

CRIME

Murder is always a mistake ... one should never do anything that one cannot talk about after dinner.

Oscar Wilde,* The Picture of Dorian Gray, *1891

I don't think you can make a lawyer honest by an act of legislature. You've got to work on his conscience. And his lack of conscience is what makes him a lawyer.

Will Rogers, 1927

***

Number of known victims of Jack the Ripper: **5**

Number of people in US jails or prisons, on probation or on parole in 2008:
7.3 million

Proportion of the world's lawyers who live in the USA:
67 per cent

Year that FBI agents were first allowed to carry guns:
1934

Number of vehicles stolen every hour in Australia: **10**
proportion of cars that are fitted with immobilisers:
45 per cent
proportion of stolen cars that have immobilisers: **7 per cent**

Proportion of children in Australia who are regularly bullied at school:
20 per cent

Number of murders in Australia in 2008: **260**
proportion finalised within thirty days of a victim becoming known to police: **77.7 per cent**

Proportion of defendants in higher courts in Australia who, in 2005–6, were:
proven guilty: **79 per cent**
acquitted: **8 per cent**

Proportion of defendants in magistrate's courts in Australia who, in 2005–6, were:
proven guilty: **85 per cent**
acquitted: **4 per cent**

Proportion of school bullies in Australia who reportedly have a criminal record by age 30:
25 per cent

Year that whipping young children was banned in Britain:
1932

Recorded number of women to have been tortured in the Tower of London:
1 (Anne Askew for heresy in 1545)

World's longest criminal trial (for murder, in Hong Kong between 30 November 1992 and 29 November 1994):
398 days

number of defendants: **14**
number found guilty: **0**

Year the FBI was founded in the USA:
1908

I think crime pays a lot. The hours are good, you travel a lot.

***Woody Allen,* Take the Money and Run,**
screenplay,1969

***

Proportion of current surveyed Australian drivers aged under 25 who believe taking drugs (such as cannabis, speed, cocaine, ecstasy) does not affect their driving:
13 per cent
in 2001: **5 per cent**

Number of cases before magistrates courts in Australia in 2008:
about 530,000
age group of those most likely to receive a custodial sentence:
25–34 years

Number of people (sentenced and unsentenced) in adult prison in Australia as at 30 June 2009:
23,900
proportion who had served time in prison previously:
56 per cent
median age:
32 years

Proportion of applications to the High Court of Australia that involved unrepresented litigants:
in 2004–5: **58 per cent**
in 1995–6: **20 per cent**

Proportion of murders in Australia that are caused by:
a stab wound: **32 per cent**
a beating: **27 per cent**
a gunshot wound: **17 per cent**
strangulation/suffocation: **12 per cent**
another cause: **12 per cent**

Proportion of surveyed Australian women who had experienced an incident of violence in the previous 12 months:
aged 18–24: **19 per cent**
aged 35–44: **6.8 per cent**
aged over 55: **1.2 per cent**

Year that Scotsman Allan Pinkerton opened the world's first private detective agency in Chicago:
1850

Survival time of US President McKinley after being shot by an anarchist in 1901:
9 days

Proportion of those charged with illicit drug offences proven guilty in Australian magistrates courts: **99 per cent** in higher courts: **97 per cent**

Number of fraud offences committed in Australia each year:
about 105,000

I haven't committed a crime. What I did was fail to comply with the law.

David Dinkins,* Mayor of New York, at suggestions of tax evasion, *1993

***

Prisoner rate per 100,000 population in the Northern Territory
(Australia's highest): **514**
in Victoria (lowest): **98**

Proportion of burglaries in the USA committed by people aged 13–21:
80 per cent

Annual cost of alcohol and drug related crime to the Australian community:

$2 billion–$4 billion

Proportion of Australian businesses that have been the victim of online credit card fraud:
32 per cent

Proportion of robbery victims in Australia in 2008 that were male: **74 per cent**
of blackmail/extortion victims: **65 per cent**
of attempted murder victims: **72 per cent**
of murder victims: **62 per cent**
of kidnapping/abduction victim: **39 per cent**

Proportion of motor vehicle thefts in Australia that are from a community location:
54 per cent
residential location:
26 per cent
other location:
20 per cent

Amount stolen in the 1963 Great Train Robbery in the UK:
£2.6 million
its value today:
£32 million ($A57 million)
total years of imprisonment received by the gang:
307

Year the polygraph was invented (by John A. Larson, USA):
1921

Once in the racket you're always in it.

Al Capone, 1929

***

Percentage of prisoners in the USA who apologise for their crime in their last words before execution: **30 per cent**

Number of murders for which gangster John Dillinger was wanted when he was shot down outside a cinema in Chicago,
USA: **16**
year: **1934**

Number of charges relating to prostitution rackets for which Lucky Luciano was found guilty in 1936: **62**
number of women he employed: **1,000**
number of houses: **200**

Number of judges in the original High Court of Australia: **3**
year: **1903**

Number of Nazis sentenced to death at the 1946 Nuremberg Trials: **12**
to life imprisonment: **1 (Rudolf Hess)**
jailed: **5**
acquitted: **3**

In Florida, USA, the average age of a death-row inmate: **37**
average age at the time of the offence: **30**

Year the last person was hanged in Australia (Ronald Ryan, Melbourne's Pentridge Prison):
1967

Number of manhole covers stolen from city streets during a six-week period in Shanghai, China in 2004: **1,826**
number of people who were killed when they fell down the open holes during this time: **2**

Number of Japanese midget submarines that attacked Sydney Harbour in May 1942: **3**

Year that bushranger Ned Kelly was hanged: **1880**

Proportion of crimes in Australia that are reported to police:
motor vehicle thefts: **90 per cent**
break-ins: **75 per cent**
assaults: **30 per cent**

Proportion of prisoners in Australia at 30 June 2009 that are aged between 20 – 39 years: **67 per cent**

Proportion of robberies in Australia that involve a weapon:

36 per cent

Proportion of prison escapees in the USA that are recaptured:
94 per cent

SCIENCE AND SPACE

Sattinger's Law: It works better if you plug it in.

Arthur Bloch,* Murphy's Law and Other Reasons Why Things Go Wrong, *1977

I have picked up a pretty sound working knowledge of electrical matters. It is not comprehensive, God knows– I still can't fully fathom why you can't boil an egg on an electric guitar.

Keith Waterhouse,* The Passing of the Third-floor Buck, *1974

***

Year that the last astronaut landed on the Moon (Gene Cernan):
1972

Estimated number of stars in the Milky Way galaxy:
over 100 billion
number we can see with the naked eye from Earth:
about 3,000

Hearing range of a human: **20 hertz to 20,000 hertz**
of a rat: **1,000 hertz to 50,000 hertz**

Weight of the average smell:
760 nanograms

Number of planets that rotate clockwise:
1 (Venus)

Mass of the moon:
7.35×10^{22} kilograms

as a fraction of the Earth:
1.23 per cent

Average weight of an iceberg:
20 million tonnes
proportion below water:
90 per cent

Time it takes the Sun's rays to reach the Earth:
about 8 minutes

Average number of lightning strikes to the Earth each second:
100

Estimated age of the universe:
13.7 billion years

Temperature at which degrees Centigrade and Fahrenheit are equal:
minus 40

Time it takes sunlight to reach Pluto:
over 5 hours

Number of degrees in each angle of an equilateral triangle: **60**

Number of Earth days in a Martian year:
669.774

Average intelligible outdoor range of a male human voice in still air:
180 metres

To err is human, but to really foul things up requires a computer.

Anon.

***

Distance in a light year:
9.46 trillion kilometres

Estimated temperature at the centre of the Sun:
15.6 billion degrees Centigrade

Typical length of a lightening bolt:
9 kilometres

Average distance from the Earth to the Moon:
384,403 kilometres

Temperature in the Boomerang Nebula (the coldest place in the Milky Way): **minus 272 degrees Centigrade**
distance from Earth: **about 5,000 light years**

Closest approach to the Sun by a spacecraft (*Helios 2* in 1976):
43.5 million kilometres

Speed at which objects travel as they approach a black hole:

up to 32 million kilometres per hour

Average surface temperature of the Earth:
15°C

Weight of a person weighing 65 kilograms on Earth:
on the moon: **11 kilograms**
on Pluto: **3.25 kilograms**
on Jupiter: **165 kilograms**

Largest known prime number in May 2010: $\mathbf{2^{43112609}-1}$
number of digits: **12,978,189**
year of discovery: **2008**

Number of litres in a barrel of oil:
159

Number of litres in a gigalitre:
1,000 million
approximate number of Olympic size swimming pools in a gigalitre:
400–500

Number of people who have set foot on the moon:
12

Year the *Voyager I* spacecraft was launched from Earth: **1977**
distance travelled so far: **about 17 billion kilometres**

The billionth digit of the number *pi:* **9**

At the Earth's surface, number of times the Moon's gravitational force is greater than that of the Sun: **2.2**

Well-informed people know that it is impossible to transmit the voice over wires and, if it were possible to do so, the thing would be of no practical value.

Boston Post, 1865

***

Surface area of the Earth:
about 510,000,000 square kilometres

Air pressure per square metre at ground level:
10 tonnes

Melting point of aluminium: **660.32°C**
boiling point: **2519°C**

Length of a nanosecond (*ns* or *nsec*):
1 billionth (10^{9}) of a second

Proportion of the total weight of the solar system that is due to the Sun:
99.87 per cent

Proportion of the world's freshwater runoff used by humans:
60 per cent

Proportion of body heat lost by not covering your head when it is cold:

50–70 per cent

Number of days in the rotation of the planet Venus:
243

Number of orbits of the Earth made by the first woman in space (Valentina V. Tereshkova, USSR, on a three-day mission in 1963): **48**

Thickness of the ice that covers Antarctica:
about 2,000 metres

Estimated temperature of the Earth's core:
3000–7000°C

Speed of sound in dry air at sea level:
332 metres/second
in fresh water:
1,410 metres/second

Diameter of Pluto:
about 2,300 kilometres

Distance of the closest major galaxy from Earth (Andromeda):
2.3 million light years

Time of the longest-lasting recorded rainbow (in Britain, 1994):
6 hours

Number of hydrogen atoms in a molecule of water:
2

Space isn't remote at all. It's only an hour's drive away if your car could go straight upwards.

Fred Hoyle* in Observer, *9 September 1979

***

Number of known asteroids (most between Jupiter and Mars):
more than 1 million
number with Earth-crossing orbits:
more than 1,000
proportion orbiting between Jupiter and Mars:
95 per cent

Number of grams in a dram:
1.77195

Mass of the Earth:
about 6,000 trillion tonnes

Number of cells in a human brain:
about 10,000 million

Youngest person to have a pacemaker (the size of a postage stamp) fitted:
4 hours old

Proportion of DNA that humans share with chimpanzees:

99 per cent

Average surface temperature of Venus:
480 degrees Centigrade

Longest recorded case of survival without food or water:
18 days

Rotational speed of the Earth at the Equator:
1,720 kilometres per hour

Reported number of lines of code in Microsoft Windows XP:
about 40 million
Vista: **50 million**

Distance between the highest mountain on Earth and the deepest point in the ocean:
19 kilometres

Number of times the mass of Jupiter is to that of the Earth:
318

Year that the Bunsen Burner was invented (by Robert Wilhelm Bunsen):
1855

Proportion of the Earth’s land surface that is permanently covered in ice:
10–15 per cent

All science is either physics or stamp collecting.

Ernest Rutherford (1871–1937)

***

Year that Thomas Edison took out his patent, in the USA, for the electric lamp:
1879

Mass of the Sun: **1.989×10^{30} kilograms**
times it is greater than the mass of the Earth:
330,000

Rotation period of the Sun in Earth days:
25–36

Escape velocity from:
the Earth: **11.2 kilometres/second**
the Moon: **2.4 kilometres/second**

Average distance from Mars to the Sun:
227.9 million kilometres

Temperature of absolute zero:
minus 273.15° Centigrade

Proportion of brass that is copper: **55 per cent**
zinc: **45 per cent**

Number of seconds in a (non–leap) year:
31,536,000

Estimated time the Sun will continue to burn:
5–6 billion years

Proportion of steel in a tin can:
about 99 per cent

Number of atoms in a drop of water:
10^{14}

Estimated number of stars in the universe:
70,000 million million

Longest spacewalk (by Jim Voss and Susan Helms in 2001):

8 hours 56 minutes

Space is almost infinite. As a matter of fact, we think it is infinite.

Dan Quayle,* in the Daily Telegraph, *1989

***

Maximum temperature on Mars: **25°C**
minimum: **minus 125°C**

Proportion of dry air that is:
nitrogen: **78.08 per cent**
oxygen: **20.95 per cent**
argon: **0.93 per cent**
all other gasses: **0.04 per cent**

Average temperature on the planet Neptune: **minus 218°C**
number of Earth years it takes to orbit the Sun: **164.8**
length of its winter: **41.2 years**

Circumference of the Earth at the equator: **40,076 kilometres**
diameter: **12,756 kilometres**

Speed of an avalanche:
up to 350 kilometres per hour

Typical lifetime of a cumulus cloud:
10–15 minutes

Oldest recorded meteorite impact on Earth: **3.47 billion years**
estimated width: **20 kilometres**

HOME AND GARDEN

What is a weed? A plant whose virtues have not been discovered.

Ralph Waldo Emerson,* Fortune of the Republic, *1878

All really keen gardeners possess a keen sense of humus.

W.C. Sellar and R.J. Yeatman,* Garden Rubbish, *1930

***

Proportion of women who say they are extremely impressed when their partner does the housework:
86 per cent
remembers birthdays and anniversaries:
68 per cent

Proportion of surveyed British women who say they get more pleasure from cleaning the house than having sex:
25 per cent
of those aged under 35:
40 per cent

Size of the average Australian home:
in 2009: **215 square metres**
in 1984: **150 square metres**

Number of British schoolchildren educated at home:
about 150,000

Proportion of Americans who would like to die at home:

70 per cent
proportion who actually do so: **25 per cent**

Proportion of house dust that is dead skin cells:
70 per cent

People aged in their twenties in Australia who still live with their parents: **30 per cent**
in 1976: **21 per cent**

Proportion of males aged 18–30 in Italy who live with their parents: **80 per cent**
in the USA: **43 per cent**

Proportion of surveyed British women who only clean the house if guests are coming to stay:
60 per cent
who fake cleaning the house by stuffing objects into cupboards and drawers:
26 per cent

Proportion of three-bedroom dwellings in Australia that have only one person living in them: **20 per cent**
two people: **38 per cent**

Average time spent by Japanese watching television each day (world's highest):
4 hours 30 minutes

Proportion of the world's rose species that come from Asia:
80 per cent

Number of species of begonia: **around 1,000**
number of hybrids: **1,000–2,000**

Hatred of domestic work is a natural and admirable result of civilisation.

Rebecca West,* The Freewoman, *6 June 1912

***

Average size of households in Australia:
in 2008: **2.53**
in 1995: **2.69**
in 1911: **4.50**

Proportion of Australian households that recycle waste:
95 per cent

Proportion of surveyed Australians who have a high level of trust in their neighbours:
33 per cent
in members of their immediate family:
90 per cent

Proportion of dwellings in Australia that have access to a public sewerage system:
73 per cent

Proportion of households in Australia with:
an automobile: **85 per cent**
a telephone: **95 per cent**
a washing machine: **90 per cent**

Proportion of an earthworm that is protein:
72 per cent

Number of species of ferns: **10,000**
algae: **12,000**
fungi: **100,000**

Maximum daily growth rate of bamboo:
30–60 centimetres

Rate of growth of the fastest growing weed (Japanese Knotweed): **10 centimetres per day**
depth of roots: **2 metres**
height: **1–3 metres**

Maximum height of a sunflower: **8 metres**
diameter of the corolla: **80 centimetres**

Average weight of a coconut: **8 kilograms**
maximum weight: **15 kilograms**

Length of a petal on Jacob's Ladder orchid:
25–30 centimetres

Height of the Giant Saguaro cactus found in the USA and Mexico:
up to 16 metres

I haven't reported my missing credit card to the police because whoever stole it is spending less than my wife.

Ilie Nastase

***

Proportion of oxygen required of all living things on Earth that is provided by algae:
70 per cent

Number of identifiable varieties of roses in the world:
about 13,000

Estimated number of flowering plant species in the world:
about 260,000

Proportion of a raw apple that is water:
84 per cent

Height to which cotton grows:
up to 3.5 metres

Time it takes an edible mushroom to grow to its full size:
7 days

Estimated number of species of tree:

about 100,000

FOOD AND BEVERAGE

Never drink black coffee at lunch; it will keep you awake in the afternoon.

Jilly Cooper,* How to Survive from Nine to Five, *1970

Food is an important part of a balanced diet.

Fran Lebowitz,* Metropolitan Life, *1978

***

Average weekly household expenditure in Australia on:
food and non-alcoholic beverages: **$153**
transport: **$139**
housing costs: **$135**

Estimated annual amount that households in Australia spend on goods and services they never use:
$10 billion
proportion that is due to uneaten food:
55 per cent

Amount the average British woman spends annually trying to slim:
$A2800

Proportion of teenage girls in Singapore who give in to their food cravings then go on a diet:
58 per cent

Proportion of divorce petitions in the UK that feature alcohol:
33 per cent

Number of bananas in the world's largest bunch (in the Canary Islands, 2001): **473**
their weight: **130 kilograms**

Number of varieties of apple: **over 7,000**
total weight of apples produced worldwide each year: **40 million tonnes**

Number of cafes and restaurants in Australia: **about 15,000**
annual income generated: **over $10 billion**
proportion that are licensed: **34 per cent**
seats available for dining: **1.1 million**
employees: **about 190,000**
proportion of these who are casual: **53 per cent**

World's heaviest apple (grown and picked by Chisato Iwasaki Linton in Japan, 2005):
1.849 kilograms

One reason I don't drink is that I want to know if I'm having a good time.

Nancy Astor

***

Proportion of Australian women not drinking enough fluid to hydrate their bodies:
80 per cent

Proportion of Australian adults who:
do not regularly eat vegetables: **70 per cent**
do not eat enough fruit: **48 per cent**
are physically inactive: **32 per cent**

Proportion of surveyed Australians who say they are addicted to:
coffee: **33 per cent**
chocolate: **29 per cent**
soft drinks: **23 per cent**
lollies: **20 per cent**

Proportion of Australians who occasionally eat a takeaway meal for breakfast: **35 per cent**
always: **2 per cent**

Proportion of Australians who have had a full glass of alcohol by age 14:
9 per cent

Number of meat pies sold in Australia each year:
about 250 million

Proportion of Australian males aged 14–17 who are regular weekly drinkers: **20 per cent**
of females: **17 per cent**

Number of surveyed Australians who say they are on a low carbohydrate diet: **2.5 million**
proportion of all adults: **20 per cent**

Proportion of Australians who occasionally eat a takeaway meal for lunch: **92 per cent**
always: **22 per cent**

Average annual intake of caffeine per capita in Finland (world's highest):
145 grams

Number of bottles of champagne consumed in 2008:
in France: **181 million**
in the UK: **36 million**
in the USA: **17 million**

Reported number of pieces of fruit and vegetables thrown away by Britons each week:
97 million
annual cost:
$A3.75 billion

Fastest drinking of a pint (568 ml) of Guinness stout (by Peter Dowdeswell, UK, 2001):
2.1 seconds

I must get out of these wet clothes and into a dry martini.

Alexander Woolcott

***

Amount Britons spend on chocolate each year:
$A7 million

Average amount of cakes and biscuits eaten each week by the average Briton: **320 grams**
of fresh fruit: **229 grams**

Proportion of the world's coconuts that are grown in the Asia-Pacific region:
93 per cent

Estimated number of vegetarians in India:
220 million

Proportion of Americans who drink coffee:
55 per cent

Proportion of households in the UK with at least one vegetarian:
6 per cent

Amount of food Australians throw away each year:
3.3 million tonnes

Average number of eggs eaten by Americans each year:
286

Proportion of Australians who:
believe that getting drunk occasionally is part of being Australian: **50 per cent**
feel pressured to drink alcohol in a workplace situation:
17 per cent
know someone with an alcohol problem: **33 per cent**

Amount paid at a London auction in February 2006 for a bottle of vintage Chateau d'Yquem:
$A130,000
year it was bottled:
1787

Production of red/rose wine in Australia in 2008–9:
623.1 million litres
white wine:
537.3 million litres

Amount of beer in a 'pony':
140 millilitres

Heaviest ever pumpkin (grown by Christy Harp, USA, 2009):
784 kilograms

Proportion of Britons who take their own sweets to the cinema:
50 per cent

Number of olive trees in the world:
about 800 million

Once, during Prohibition, I was forced to live for days on nothing but food and water.

W.C. Fields

***

Year the first drinkable man-made glass of carbonated water was created (by Joseph Priestley):
1796

Proportion of surveyed Australians who nominated as a favourite thing to drink:
wine: **28 per cent**
any soft drink: **27 per cent**
water/mineral water: **27 per cent**
beer: **21 per cent**
tea/iced tea: **13 per cent**
juice: **13 per cent**
coffee: **10 per cent**

Proportion of Australians who nominated as a favourite thing to eat:
vegetables: **11 per cent**
pasta: **9 per cent**
roast meat: **8 per cent**
chicken: **8 per cent**
fruit: **7 per cent**
steak: **7 per cent**

Proportion of full-fat milk that is water:
87 per cent

Amount of peanuts and peanut butter eaten by the average American each year:
3 kilograms

Proportion of shoppers who are too embarrassed to put budget items in their trolley:
18 per cent

Year the paper drinking straw was patented (by Marvin C. Stone, USA): **1880**
his ideal length for them: **21.5 centimetres**

Number of cups of coffee drunk around the world each day:
400 million–1 billion

Number of grapes produced annually worldwide:
70 million tonnes
number of varieties:
about 6,000

Number of doughnuts made in the USA each year:
10 billion

Proportion of the world's almonds that are used in chocolate bars:
40 per cent

Volume of a 'drop': **0.08 millilitres**
a 'dash': **0.62 millilitres**

I'm President of the United States and I'm not going to eat any more broccoli.

George H.W. Bush,* New York Times, *1990

***

Average weight of an ostrich egg: **1.36 kilograms**
time it takes to hard boil it: **40 minutes**

Amount of wine exported from Australia in 2007-08:
698 million litres
its value:
about $2.5 billion

Proportion of the world's caviar that comes from the Caspian Sea:
90 per cent

Proportion of caviar that is protein: **25 per cent**
fat: **19 per cent**

Year that ginger ale was created (by John McLaughlin, Canada):
1904

Proportion of British men who when they wake up:
want a cup of tea: **42 per cent**
who want sex: **5 per cent**

Number of Russians who die each year from alcoholic poisoning:
about 40,000

Proportion of Australia women who say they like the taste of beer:
37 per cent
who believe that beer is a 'masculine' drink:
42 per cent

Number of glazed doughnuts eaten in eight minutes (by Eric Booker, New York): **49**

Most empty beer kegs balanced on a head (for 10 seconds by John Evans, UK, in 1998): **11**

Year of the Boston Tea Party: **1723**

Common name for the original Bib-Label Lithiated Lemon Sodas:
7UP

Average amount of beef eaten by Australians each year:
36 kilograms

Number of pancakes tossed in two minutes by Ralf Laue of
Germany in 1997: **416**

Year that Grand Marnier was created (by Alexandre Marnier-Lapostolle for the Prince of Wales):
1880

Volume of beer drunk worldwide each year:
about 140 billion litres

I've made it a rule never to drink by daylight and never to refuse a drink after dark.

H.L. Mencken,* New York Post, *1945

***

The only edible food for humans that will never go bad:
1 (honey)

Amount of cheesecake eaten in nine minutes by Sonya Thomas:
4.99 kilograms
her weight at the time:
47.6 kilograms

Proportion of a cabbage that is water:
91 per cent

Estimated bubbles in a bottle of champagne:
49 million

Volume of a 10 gallon hat:
about 2.8 litres

Amount of bacon produced in the US each year:
about 900,000 tons

Year Coca Cola was invented (by John Pemberton, a pharmacist):
1886

MONEY AND BUSINESS

I don't want money. It is only people who pay bills who want that, and I never pay mine.

Oscar Wilde,* The Picture of Dorian Gray, *1891

You can't help liking the boss. If you don't, they fire you.

Anon.

***

Proportion of Americans who say they would continue to work if they won $10 million in the lottery:
55 per cent

Cost of the average wedding in the USA in 2009:
$A30, 860

Average amount of cash an American has in their wallet:
$104

Number of credit and charge cards held by Australians in 2009:
15 million
total amount owing on them:
$41 billion

Number of different ways to make change for a dollar:
293

Number of accommodation businesses in Australia:
about 5,700
proportion that are motels: **38 per cent**

caravan parks: **20 per cent**

Proportion of people in Australia who feel loyal to their:
bank: **52 per cent**
tea or coffee brand: **49 per cent**
phone company or internet service provider: **48 per cent**
make of car: **36 per cent**
brand of washing powder: **32 per cent**
brand of petrol: **20 per cent**

Number of online shoppers in Australia in 2006: **5.9 million**
in 2004: **1.2 million**

Proportion of surveyed Australian workers in 2009 who say they have been bullied:
40 per cent

Proportion of employees in Australia who have superannuation provided by their current employer:
90 per cent

Proportion of Australians who have confidence in the organisation they work for:
40 per cent
who trust the leader:
42 per cent

Proportion of Australians who say they would recommend their job to somebody else:
44 per cent
who think their boss treats them fairly:
90 per cent

A rich man is nothing but a poor man with money.

W.C. Fields

***

Proportion of Australian workers who say they understand what human resources management does:
75 per cent
who believe it contributes to the organisation's success:
64 per cent

Proportion of Australian males who say their job has a bad effect on their sex life:
42 per cent
of Australian females:

47 per cent

Estimated annual revenue lost to an employer (in sick days and poor productivity) in Australia by a worker with untreated depression:
$9,660

Amount left in her will by Marianne White to a cemetery
in Germany:
$10 million

Proportion of Britain's wealth that is owned by the wealthiest 1 per cent:
25 per cent

Proportion of Britain's personal wealth that is owned by women:
48 per cent

Number of female millionaires aged 18–44 in Britain: **47,355**
aged over 65: **71,369**

Number of millionaires in Britain in 2009: **242,000**
in 2007: **489,000**

Average amount British household spends on transport each week:
£60

Weekly earnings of Toto (the dog) while filming the 1939 movie *The Wizard of Oz:*
$US125

He is a distinguished man of letters. He works for the Post Office.

Max Kauffmann

***

Worth of the world's pharmaceutical market in 2010:
about $A825 billion
proportion due to the USA alone:
48 per cent

Proportion of spouses in the USA who keep money hidden from their partner:
60 per cent

Maximum allowable working time per week in a coal mine for boys aged 13–18 years in NSW, Australia in 1876:
50.5 hours

Number of registered businesses in Australia:
about 3 million
proportion who do not employ anybody:
72 per cent

Proportion of Australians who feel as if they have never had a holiday by the end of their first week back at work:
38 per cent

Amount that Australians spend each year on audited magazines:
about $1 billion

Proportion of surveyed Germans who rated having a good qualification as their most important aim in life (ahead of love, money and having children):
90 per cent

Proportion of people who reportedly live on $A3 a day or less:
in Cambodia: **75 per cent**
in Papua New Guinea: **70 per cent**
in Vietnam: **47 per cent**
in China: **29 per cent**
in Thailand: **22 per cent**

Average percentage by which women in the UK earn less than their male colleagues:
27 per cent

Proportion of those who experience conflict between work and home life who want to work more hours:
67 per cent

Average amount Australians spent in 2009 on their credit cards:
$14,310

Amount paid at auction in London for three paintings done by an ape, named Congo, in 2005:
US$25,000

Total prize pool of the 2009 Spanish (*El Gordo*) Lottery:
$A3.7 billion (2.3 billion Euros)

Money can't buy friends, but you can get a better class of enemy.

Spike Milligan,* Puckoon, *1963

***

Number of Australians who will retire in the next 10 years:
over 4 million
average age at retirement:
57 years

Average length of a worker's smoke break: **5 minutes**
coffee break: **10 minutes**

Estimated number of emails sent worldwide each day:
in 2009: **250 billion**
in 2006: **135 billion**

Proportion of Australian women who are turned on by men with financial security:
7 per cent
who are turned off by men without it:
22 per cent

Proportion of surveyed Australian workers who say they intend to find a new job in the next twelve months:
70 per cent

age of most disgruntled employees:
45–54 years

Estimated amount Britons spend on car-boot sales each year:
about $A3.6 billion

Number of casinos in Australia: **13**
number of employees: **over 19,000**
proportion of these aged under 45: **65 per cent**

Proportion of surveyed Australian women who say that beauty advertisements make them feel bad about the way they look:
49 per cent

Amount Australians reportedly owe to family and friends:
$12 billion

The highest denomination bank note ever issued by the US Federal Reserve:
$100,000

Equivalent price of odds of 2–1:
$3.00

Cost then of a first-class return trip from Melbourne to England on then P & O Liner *Mooltan* in 1932:
$344

tourist class: **$140**

Proportion of Australian workers who are overqualified for their job:
27 per cent

I don't hire anybody not brighter than I am. If they're not brighter than I am, I don't need them.

Bear Bryant

***

Number of German marks to one British pound in:
July 1914: **20**
January 1921: **315**
January 1922: **932**
July 1923: **1.7 million**
August 1923: **22.5 million**
September 1923: **480.5 million**
November 1923: **20 billion**

Cost of building Disneyland in Anaheim, California:
$US17 million
year it opened: **1955**

Proportion of Australian beef production that is exported:
65 per cent
the value of exports:
$5 billion

Amount of wool produced in Australia in 2008-9:
370,000 tonnes
value: **$2.223 billion**

Proportion of sheep in Australia that are merinos:
85 per cent

Retail sales of wine in Britain in 2009:
£5.5 billion
proportion that was Australian wine:
24.3 per cent

Proportion of Australian workers aged 15–69 who have been in their current job for less than twelve months:
23 per cent

Amount spent annually by Australians on tax agents and accountants to manage their tax affairs:
over $1 billion

Level of household debt in Australia in September 2008:
$1.1 trillion
in September 1990:
$190 billion

Proportion of Australians who feel the advertising of fast food is unacceptable: **51 per cent**
of alcohol: **53 per cent**
of gambling: **79 per cent**

Year the Commonwealth Bank of Australia began as a savings bank: **1912** commercial bank: **1913**

Cost of the first Holden car produced in Australia: **£760**
year: **1948**

Money, it turned out, was exactly like sex. You thought of nothing else if you didn't have it and thought of other things if you did.

James Baldwin, 1924–1987

***

Total number of hours worked in Australia in:
July 2009: **1,516.8 million**
July 2006: **1,480.2 million**

Average number of sick days taken by Australians each year:
8.5
cost per employee: **$2,500**

Price of a litre of petrol in Australia in 1948:
14 cents

Amount of money inherited by Australians in 2003:
$14.8 billion
average inheritance:
$64,000
proportion that was less than $20,000:
50 per cent
proportion of adults who receive an inheritance in any one year:
1.4 per cent

Value of comic sales in Hong Kong each year:
about $A100 million

Amount fetched at a 2004 auction for an acoustic guitar made for singer Johnny Cash:
$A189,000

Amount fetched at a 2005 auction for a saddle used by jockey Jim Pike on Pharlap in 1930
$43,145
racing silks: **$67,000**

Amount fetched at a 2005 auction in California for the Smith & Wesson gun that killed outlaw Jesse James: **$A490,000**
year it was made: **1875**
calibre: **.44**

Proportion of managers who say that a week without email is more traumatic than getting divorced:
33 per cent

Year that the Australian Federal Government abolished tertiary education fees: **1974**
year that tertiary undergraduate fees were re-introduced: **1987**

Amount paid in 2005 to a 30-year-old American woman for having a casino website permanently tattooed on her forehead for advertising purposes: **$A10,000**

Number of billionaires ($US) in the world:
in 2010: **1,010**
in 2009: **793**
in 2008: **1,125**

When you don't have any money, the problem is food.
When you have money, the problem is sex.
When you have both, it's health.

J.P. Delaney,* The Ginger Man, *1955

***

Longest time a coin has been spun (by Scott Day, UK, 2003) before coming to a complete rest:
19.37 seconds

Year the words 'In God We Trust' first appeared on paper money in the USA (on One Dollar Silver Certificates):
1957

Most zeros in a banknote printed on a sovereign nation's currency (Yugoslav Dinara):
500,000,000,000
year of issue: **1993**

Year the Basic Wage was introduced into Australia:
1907
the amount: **£2 2/-**

Year the $2 coin replaced the $2 bank note in Australia:
1988

Gross Domestic Product (GDP) in $US per capita in 2009:
Singapore: **$50,300**
Australia: **$38,800**
Zimbabwe (lowest in the world): **less than $100**

BIRTH, DEATH, MARRIAGE AND DIVORCE

Either he's dead or my watch has stopped.

Groucho Marx

Adam and Eve had an ideal marriage. He didn't have to hear about all the men she could have married, and she didn't have to hear about the way his mother cooked.

Kimberly Broyles

***

Proportion of surveyed cohabiting couples in Australia who have definite plans to marry:
32 per cent
who end up marrying within four to seven years:
40 per cent

Number of divorces in the UK in 2007: **144,220**
in 1993: **180,018**

Proportion of couples in Australia who once had a de facto relationship now reporting they have a loveless marriage:
10 per cent
who did not have a de facto relationship:
5 per cent

Proportion of marriages that survive news of an affair:
50 per cent

Proportion of divorces in Australia that are due to infidelity:

20 per cent

Average number of marriage licences issued in Las Vegas each day in 2008:
335

Proportion of married men in Australia who are seeking a 'short-term relationship':
25 per cent
of married women:
10 per cent

The chance of having identical twins:
1 in 285

Proportion of marriages in Australia in 2010 that are performed by civil celebrants:
65 per cent

Proportion of married people in Australia who say their partner treats them very well: **43 per cent**
of de facto couples: **21 per cent**

Proportion of Australians who lived together before marriage in 2010:
76 per cent
in 1975: **16 per cent**

Median age at which Australian males in 2008 married for the first time: **29.6 years** females: **27.7 years**

Number of couples who married in China in 2009:
about 11.5 million
who divorced: **2.1 million**

Expected ratio of male births to female births per 1000 births:
511 to 489

The difference between a divorce and a legal separation is that a legal separation gives a husband time to hide his money.

Johnny Carson

***

Number of marriages in Japan in 2008: **726,000**
number of divorces: **251,000**

Average length of an engagement in the USA in 2009: **17 months**

Number of living multiple births worldwide:
over 125 million

Number of births registered in Australia in 2008 (highest on record):
296,600
median age of mothers giving birth:
30.7 years
median age of the fathers:
33.1years

Proportion of marriages in Australia in which neither party has been previously married:
67 per cent
both parties have been:
15 per cent

Number of registered male deaths in Australia in 2008: **73,500**
female: **70,400**

Proportion of women in Australia aged 25–29 who have not given birth: **59 per cent**
in 1982: **35 per cent**

Proportion of births in Australia that are premature (born less than 37 weeks gestation):
7 per cent

Number of multiple births registered in Australia in 2008:
4,600
proportion of these that were twins:
about 95 per cent

Number of marriages in the US each year where the relationship began on a coffee break:
about 10,000

Marriage is a lottery, but you can't tear up your ticket if you lose.

F.M. Knowles

***

Proportion of divorces in the UK that come from marriages of less than five years:
25 per cent

Number of US presidents who died on 4 July 1826:
2 (John Adams and Thomas Jefferson)

Year the composer Frédéric François Chopin died:
1849

Number of infant deaths per 1000 live births in Australia:
in 2009: **4.7**
in 1994: **9.2**

Number of registered marriages in Australia:
in 2008: **118,756**
in 1984: **108,655**

Proportion of marriages in Australia where the groom is older than the bride:
66 per cent

Proportion of births in Australia in 2008 whose paternity was not acknowledged: **3 per cent**
in 1988: **5 per cent**

Number of women in Britain aged over 40 who gave birth:
in 2005: **22,246**
in 1995: **11,319**

Proportion of births in Britain in 2009 that were to unmarried parents: **43 per cent**
in 1995: **33 per cent**

Estimated number of marriage matchmaker organisations in Japan:
4,000
annual value of the industry:
about $A800 million

Proportion of births in Australia that are to married women:
67 per cent
average time after marriage for their first baby:
2.5 years

Proportion of college-educated women in the USA who kept their own name after marriage:
in 2000: **17 per cent**
in 1990: **23 per cent**

Average time between births in 2008:
in the USA: **7 seconds**
in Australia: **106 seconds**

Proportion of births in Britain that are by caesarean section:
23 per cent

Death is nature's way of saying your table's ready.

Robin Williams

***

For Australian men who divorce the most likely age at which they married: **27**
for women: **24**
ages at which they most likely divorce: **35–39**

Proportion of surveyed Australian women who were unaware that they can become pregnant on only two days each month:
94 per cent

Number of people who die each year in the US from traffic accidents: **about 41,000**
burns: **6,000**
lightning strikes: **400**

Estimated length of time that a 'honeymoon effect' stays after marriage:
12 months
until a woman's sex drive usually diminishes after marriage:
3 weeks to 2 years

Proportion of pregnancies that result in Down syndrome:

1 in 770
haemophilia: **1 in 10,000**

Proportion of babies born with teeth:
1 in 2000

Number of road deaths in Australia in 2008: **1,464**
proportion of these who were drivers: **47 per cent**
passengers: **21 per cent**
pedestrians: **13 per cent**
motorcyclists: **17 per cent**
bicyclists: **2 per cent**

Estimated proportion of marriages in Australia today that are likely to end in divorce:
32 per cent

Proportion of all visits to Fiji that are for a wedding or honeymoon:
11 per cent

Proportion of bodies in Japan that are cremated:
99 per cent

Number of weddings in the USA in 2008:
2,298,000
proportion held in a place of worship:
50 per cent

Proportion of women in Australia who smoke during pregnancy:
15 per cent

Proportion of births in London that are to mothers born outside of Britain:
50 per cent

I don't believe in an afterlife, although I am bringing a change of underwear.

Woody Allen

***

Number of divorces in Australia in 2008: **47,209**
proportion that involved children: **50 per cent**
proportion of divorces filed by the wife alone: **37.3 per cent**
husband alone: **27.5 per cent**
jointly: **35.2 per cent**

Longest wedding train (Romania, April 2009):
1,579 metres

Proportion of Britons who mention their pet in their will:
10 per cent

Proportion of divorces in Australia in 2008 that were from marriages of five years or less:
25 per cent

Proportion of divorces in Britain that are instigated by women:
70 per cent

Increase in the probability that a person in the UK will divorce if they are a smoker rather than a non-smoker:
53 per cent

Proportion of babies born in Australia by caesarean section in 2007: **31 per cent**
in 1992: **18 per cent**

Average age at marriage for women in Kebbi State, Nigeria:
11 years

Average volume of saliva a baby dribbles by their first birthday:
120 litres

Australian birth rate per 100,000 population: **13.8**
world average: **21.2**

Proportion of births in the USA that take place in a vehicle:
1 in 300

Number of babies born outside marriage in Australia:
in 2008: **100,800**
in 1993: **64,789**

Average number of labour pains for first time mothers: **about 135**

Average length of an umbilical cord: **55 centimetres** typical range: **40–70 centimetres**

Proportion of Australian couples affected by infertility: **17 per cent**

Proportion of people who are cremated in Australia: **60 per cent**

Year that gangster Al Capone died: **1947**

A divorce is like an amputation. You survive it but there's less of you.

Margaret Atwood

***

Proportion of Australian couples affected by infertility:
17 per cent

Median child-bearing age of mothers in Australia:
in 2008: **31.7 years**
in 1963: **28.4 years**
in 1943: **25.7 years**

Number of people in the USA who are affected by domestic violence:
32 million

Diameter of a human eyeball:
at birth: **17–18 millimetres**
at puberty: **20–21 millimetres**
in adulthood: **24–25 millimetres**

Number of years for a ruby wedding anniversary: **40**
gold: **50**
emerald: **55**
diamond: **60**

Proportion of pregnancies for women aged under 40 years in Australia that end in a miscarriage:
20 per cent

OTHER STUFF

Never do anything yourself that others can do for you.

Agatha Christie

Wait, I just remembered something! You're boring and my legs work.

***David Spade, as Finch in* Just Shoot Me**

***

Number of taxi drivers in New York City:
about 40,000
total distance they travel each day:
about 1.6 million kilometres

Number of English letters in the Hawaiian alphabet:
12
number of these that are vowels: **5 *(a, e, i, o, u)***
Consonants: **7 *(h, k, l, m, n, p, w)***

Number of registered motor vehicles in Australia as at 31 March 2009: **15,674,436**

average age of motor vehicles in 2009: **9.9 years**
proportion older than 15 years: **20.7 per cent**

Number of categories of the Nobel Prize: **6**

Number of Pillars of Wisdom: **7**

Number of miles in a nautical mile: **1.1508**
number of kilometres: **1.852**

Proportion of Australia's overseas-born residents who are from the UK: **23 per cent**
New Zealand: **8 per cent**

Estimated number of SMS messages sent in Australia in 2008:
about 10.8 billion

Year the first modern Olympic Games was held: **1896**
ancient Olympic Games: **776BC**

Proportion of Britons who have at least four holidays a year:
5 per cent

Number of letters in the Greek alphabet: **24**

Number of portable devices (mobile phones, PDAs, computers) left in Chicago taxis each year:
about 150,000

proportion returned to their owners:
55 per cent

I find television very educating. Every time somebody turns on the set, I go into the other room and read a book.

Groucho Marx

***

Proportion of Australians with mobile phones who regularly use them for SMS messages:
85 per cent

Proportion of Australian residents who were born overseas:
23 per cent
in 1947: **10 per cent**

Proportion of males in Australia who say they always bathe between the flags: **25 per cent**
of females: **56 per cent**

Proportion of all surf rescues in Australia that are male:
90 per cent

Year the first black box flight memory recorder was invented (by Dr David Warren in Melbourne):
1958

Year that Halley's Comet will next be seen from Earth:

2061

Number of new vehicle registrations in Australia in 2009:
937,328

Proportion of Australians who say they were happy with the presents they receive at Christmas:
37 per cent
number of unwanted gifts received:
22 million
their value:
$685 million

Number of official languages of Belgium:
3 (Dutch, French, German)

Number of mobile phones per 100 inhabitants in Estonia in 2008:
190

Total distance travelled by Britons in the UK each year:
about 725 billion kilometres
proportion that is by buses or trains:
12 per cent

Number of cigarettes reportedly smuggled into Hong Kong each year:
over 128 million

Maximum speed of the *Titanic:*
21–23 knots

Heartbeat of a mouse:
600–700 beats per minute

A bank is a place where they lend you an umbrella in fair weather and ask for it back when it begins to rain.

Robert Frost

***

Number of acres in a hectare:
2.47105

Number of pence in an old Australian shilling: **12**

Number of Australians who have reached the rank of Field Marshall:
1 (Sir Thomas Blamey)

Number of umbrellas lost each year on the London transit system:
about 25,000

Number of Internet users in 2009 in:
USA: **228 million**
Japan **96 million**
India: **81 million**
Indonesia: **30 million**
worldwide: **1.734 billion**

Proportion of Australians aged under 30 who consider having a suntan as very important:

10 per cent
who try to maintain a suntan at least on a fortnightly basis:
20 per cent

Estimated number of Internet users in China in 2009:
360 million
year it was introduced there:
1994

Number of letters in the Cambodian alphabet (longest): **74**

Reported number of miners who died at the Eureka Stockade in 1854: **22**
soldiers: **5**

Number of lives lost on *HMAS Voyager* when it was rammed by the aircraft carrier *HMAS Melbourne:* **82**
number of these who were officers (including the captain): **14**
year: **1964**

Number of 'Pachinko' parlours in Japan in July 2009:
13,000
number of people there who regularly play the game:
20 million
proportion who are women:
50 per cent

Proportion of adults who have had sex in a toilet:
Australians **31 per cent**
Americans: **70 per cent (highest)**
Thai, Vietnamese: **2 per cent (lowest)**

Average speed of London traffic today: **19 kilometres per hour**
in 1900: **19 kilometres per hour**

My wife and I were happy for twenty years. Then we met.

Rodney Dangerfield

***

Number of recognised universities in India:
about 300
of university affiliated colleges:
about 12,000

Median age of males in Malaysia in 2009: **24.3 years**
of females: **25.6 years**

Proportion of the world's silk supply produced by China:
70 per cent

Proportion of the British population who are left-handed:
12 per cent

Number of baggage items sent to departing aircraft on a busy day at Sydney International Airport: **about 19,000**
peak times: **6–10am and 12–8pm**

Estimated number of caravans and campervans in use in Australia: **330,000**

number of caravan and tourist parks:
over 2,700

Number of sides on a snowflake: **6**

Number of bits in a byte: **8**

Minimum number of steps adults need to take each day to get sufficient exercise: **10,000**

Number of votes cast in the final week of *American Idol* in the USA in 2009: **28.8 million**
in 2006: **63 million**

Time taken for French student Alex Lemaire to calculate the 13th root of a 100-digit number using mental arithmetic:
3.62 seconds
the 13th root of a 200-digit number:
513 seconds

Proportion of an egg's weight that is due to the shell:
12 per cent

Number of people in Australia studying for a qualification:
2.3 million
proportion aged 15–19 years: **26 per cent**
20–24 years: **48 per cent**

Number of verses in 'Advance Australia Fair': **5**

Number of lines in the Japanese national anthem: **4**

Bigamy is having one wife too many. Monogamy is the same.

Oscar Wilde

***

Minimum wind speed for a tropical cyclone:
119 kilometres per hour

Number of full-time students accepted into universities in the UK in 2009:
456,627
number who were mature age (over 25 years):
77,286

Width of a ten pin bowling alley: **1.04–1.07 metres**

length: **19.16 metres**

Proportion of surveyed Americans who:
attend religious ceremonies at least occasionally: **90 per cent**
who believe in God: **76 per cent**
who believe in an afterlife: **60 per cent**

Proportion of primary school teachers in Australia who are women:
80 per cent

Proportion of 55–64-year-old Australians who plan to travel the world within the next five years:
33 per cent

Proportion of adults in the USA who were regular smokers:
in 2009 **21 per cent**
in 2003: **22 per cent**
in 2002: **23 per cent**

Longest recorded human neck (women of the Padaung tribe, Myanmar):
40 centimetres

Proportion of road crashes in which fatigue is a factor:
20 per cent

Greatest distance travelled walking on hands (over 55 days in 10-hour stints from Vienna to Paris in 1900):
1,400 kilometres

The trouble with being punctual is that nobody's there to appreciate it.

Franklin P. Jones

***

Number of cigarettes sold in the USA in 2006:
371 billion
in 1982: **634 billion**

Number of items reported lost in Japan in 2008:
17.34 million
number of items returned to their owners:
6.17 million

Number of suicides in Australia:
in 2007: **1,881**
proportion who were male: **77.3 per cent**

Proportion of complaints made to British Rail that are about the way that they handle complaints:
4 per cent

Average diameter of a raindrop:
0.5 centimetres

Amount of litter removed each year from the River Thames in England:
1,000 tonnes

Number of masts on a schooner:
2 or more
Number of ways to tie a knot in a conventional tie:
85
number of these that are aesthetically pleasing: **13**

Estimated number of registered cars in the USA:
in 2009: **125 million**
in 1920: **9 million**
in 1905: **79,000**

Proportion of plane accidents that occur during:
takeoff: **35 per cent**
landing: **60 per cent**
other: **5 per cent**

Age of the world's oldest bound book (the Book of Psalms):
1600 years

Proportion of surveyed British women who think their legs are:
average: **60 per cent**
poor: **30 per cent**
perfect: **1 per cent**

Proportion of Australian university students who:
change their course mid-degree: **12 per cent**
drop out altogether before finishing: **14 per cent**

Age of the world's youngest university graduate (BA in anthropology from the University of South Alabama, USA):
10 years 4 months

I chased a girl for two years only to discover that her tastes were exactly like mine: we were both crazy about girls.

Groucho Marx

***

Number of toys sold in the US in 2009:
about 3 billion

Proportion of people who believe you should abstain from sex
until you are married:
Australians: **5 per cent**
Indonesians: **54 per cent (highest)**
Netherlanders: **1 per cent (lowest)**

Reported number of cases of Dengue fever in Singapore:
in 2009: **7,031**
in 2007: **8,826**

Number of nickels in a dime: **2**

Maximum number of clubs allowed when playing a round of golf:
14

Longest procession of Rolls-Royce cars on a public highway (in May 2000 on the A55, England):
420

Number of minutes in a day: **1,440**
in a week: **10,080**

Longest distance a human has been fired from a cannon (David Smith Snr., USA, 1999):
56.54 metres

Number of verses in Greece's nation anthem:
158

Increased chance of not being killed or seriously injured when wearing a seatbelt in a car accident:
50 per cent

Height of a limited edition Armani Barbie:
29 centimetres

Roman numerals for 2010:
MMX

Highest possible hand in a game of cribbage: **29**

Number of sides on a dodecahedron: **30**

Number in a gross: **144**

Number of dots on a normal six-sided die: **21**

Number of sides on:
a dodecagon: **12**
an icosagon: **20**

All the things I really like to do are either illegal, immoral or fattening.

Alexander Woollcott, 1887–1943

***

Number of edges on an Australian 50 cent coin: **12**

Typical number of tiles in a mah-jong set:
144

Height of the portrait *Mona Lisa:* **77 centimetres**
width: **53 centimetres**

Number of zeros in a British billion: **12**
in an American billion: **9**

Proportion of Australian drivers who say they would not pull over if they felt fatigued:
28 per cent

Average number of calories burnt from 30 minutes of playing billiards: **90**
of swimming at 3 kilometres per hour: **around 300**

Number of tricks for a small slam in bridge: **12**
grand slam: **13**

Number of months that do not have 31 days: **5**

Melting point of uranium:
1,132°C

According to the Bible, to which age Abraham lived:
175 years

Number of tiles in a game of standard Scrabble: **100**
number of these that are blank: **2**

Width of a double bed: **137 centimetres**
queen size bed: **152 centimetres**
king size bed: **183 centimetres**

Pieces of silver given to Judas: **30**

Number of psalms in the Bible:
150 chapters

Reported proportion of plane crashes that have survivors:
90 per cent

Land area of Monaco:
1.95 square kilometres

Proportion of packaging materials in Australia that are recycled:
51 per cent

Number of numbers from 1 to 99 that contain the letter ‘a’: **0**

Further Reading

Some of the many references consulted in the writing of this volume are listed below. They provide a rich source of numerical and other information.

Australian Bureau of Statistics
(URL: www.abs.gov.au)

Book of Historical Records (2000), Norris McWhirter, Virgin Publishing

Chronicles of the 20th Century (1990), Chronicle Publications, Mt. Kisko, New York

Durex Annual Global Sex Survey
(URL: www.durex.com/uk/globalsexsurvey)

Encyclopaedia Britannica Online,
(URL: www.britannica.com)

Guinness World Records, (annually), (1956 – 2010)
(URL: www.guinnessworldrecords.com)

Hot Sex: how to do it (1999), Tracey Cox, Pub. Bantam

Macquarie Book of Events (1983), ed. Bryce Weldon, Pub. Weldon

Number Crunch (since 2000), John Croucher, weekly column, *Good Weekend,* Sydney Morning Herald, Fairfax

Number Crunch (since 2000), John Croucher, weekly column, *Weekender,* Melbourne Age, Fairfax

The Big Book of Records (2003), ed. Maureen Spurgeon, Pub. Brown Watson, England

The Book of Numbers (2000), William Hartston, Metro Pub. Ltd., London

The Odd Essential Body (2005), Stephen Juan, Pub. Harper Collins, Australia

The Odd Sex (2001), Stephen Juan, Pub. Harper Collins, Australia

This Book of More Perfectly Useless Information (2005), Mitchell Symons, Pub. Transworld, UK

Vital Statistics (2005), Paul Grobman, Pub. Plume, USA

What are the Chances of That? (2004), William Hartston, Metro Pub. Ltd., London

Back Cover Material

EIGHTEEN MONTHS on the TOILET

The best of Number Crunch 1

Did you know...

The average person laughs about 15 times a day?

A woman spends an average of 8 minutes 15 seconds in a store if accompanied by another woman, and 4 minutes 41 seconds in a store if accompanied by a man?

40 per cent of British men admit that, if they had the choice, they would rather their football team win the Premiership than have sex with the woman of their dreams?

Over 29,000 rubber ducks are believed to be floating in the world's oceans?

We will each spend, on average, 9,000 hours (that would be over 16 waking hours a day for 18 months) on the toilet?

This book was the response to John S Croucher's newspaper columns that appear around the world. For over ten years the weekly

'Number Crunch' column in the *Sydney Morning Herald* and *The Age* in Melbourne has provided riveting data to Australia. These and John's other international columns are read by millions of readers each week and provide testimony to just how fascinating – and just how much fun – numbers can be. With over 2,500 facts and figures, and around 150 matching quotations, there is plenty of information to keep you absorbed and thoroughly entertained.

Books For ALL Kinds of Readers

At ReadHowYouWant we understand that one size does not fit all types of readers. Our innovative, patent pending technology allows us to design new formats to make reading easier and more enjoyable for you. This helps improve your speed of reading and your comprehension. Our EasyRead printed books have been optimized to improve word recognition, ease eye tracking by adjusting word and line spacing as well as minimizing hyphenation. Our EasyRead SuperLarge editions have been developed to make reading easier and more accessible for vision-impaired readers. We offer Braille and DAISY formats of our books and all popular E-Book formats.

Printed in Dunstable, United Kingdom

91341972R00228